PHOTOGRAPHIC ATLAS OF AN ACCRETIONARY PRISM:
GEOLOGIC STRUCTURES OF THE SHIMANTO BELT, JAPAN

Photographic Atlas
of an Accretionary Prism

Geologic Structures of the Shimanto Belt, Japan

Asahiko Taira, Timothy Byrne and Juichiro Ashi

Springer-Verlag Berlin Heidelberg GmbH

Asahiko Taira
Professor, Ocean Research Institute
University of Tokyo
1-15-1 Minamidai, Nakano-ku, Tokyo 164, Japan

Timothy Byrne
Assistant Professor, Department of Geology and
    Geophysics
The University of Connecticut
354 Mansfield Road, Storrs, CT 06269-2045, U.S.A.

Juichiro Ashi
JSPS Research Fellow, Ocean Research Institute
University of Tokyo
1-15-1 Minamidai, Nakano-ku, Tokyo 164, Japan

Publication of this book was partially supported by a Grant-in-Aid for the Publication of Scientific Research Results from the Ministry of Education, Science and
Culture of Japan.

ISBN 978-3-662-09284-2   ISBN 978-3-662-09282-8 (eBook)
DOI 10.1007/978-3-662-09282-8

PHOTOGRAPHIC ATLAS OF AN ACCRETIONARY PRISM:
GEOLOGIC STRUCTURES OF THE SHIMANTO BELT, JAPAN

CONTENTS

1 INTRODUCTION

1-1 Purpose

The sedimentary, igneous and metamorphic rocks of southwest Japan represent one of the best-documented subduction complexes in the world. Over the past several decades these rocks have been studied extensively by Japanese, French, British and American geologists, and the resulting data indicate that this continental margin has evolved as thousands of kilometers of oceanic crust were subducted over the past 100 to 200 m.y. The Cretaceous to Tertiary accretionary sequences called the Shimanto Belt are representative of the other sequences in southwest Japan and are particularly appropriate for scientific analysis and a photographic essay because they range substantially in metamorphic grade (from high temperature eclogite to zeolite facies) and are beautifully exposed along the coastal sea cliffs and deep river gorges of Shikoku Island.

In this atlas we attempt to summarize the complex history of these exceptional exposed rocks through a series of outcrop- to microscopic-scale photographs, sketches and geologic maps. The overall objective of the atlas is to broaden the base of general knowledge of the growth of continental margins and to provide an integrated framework for more detailed studies. The atlas provides both a general introduction to research in accretionary prisms and an educational aid for students in sedimentology, structural geology and tectonics. In order to put the photographs, sketches and maps in a regional setting, we have also included a brief summary of the geologic setting of southwest Japan. Most of the photographs recorded in this atlas were taken by the authors in the past 15 yrs.

In the last two chapters of the atlas, we include photographs from two very young and exceptionally well-studied accretionary prisms: the Pliocene sedimentary sequences exposed on the Miura and Boso Peninsulas and the active accretionary prism just north (or landward) of the Nankai Trough in southwest Japan. These two chapters complement the photographic essay on the Shimanto Belt and provide an important link to the modern tectonic processes that characterize active convergent plate boundaries around the world.

1-2 How to Read This Atlas

This atlas is composed of main text, figures and photographs and the figures and photographs have been quoted in appropriate places in the main text. The photographs are also grouped according to the explanation in the main text, so the readers can access the meaning of photographs easily through either text or photo captions.

2 GEOTECTONIC FRAMEWORK OF SOUTHWEST JAPAN

2-1 Geology of Southwest Japan

The present tectonic framework of southwest Japan, and Japan in general, is dominated by three plate boundaries (Figures 1 and 2). Two of these, the Nankai Trough and the Japan Trench, are relatively straightforward and reflect subduction of the Philippine and Pacific plates, respectively. The third plate boundary appears to be more complex and is represented by an incipient zone of backarc thrusting along the eastern margin of the Japan Sea, by deformation along the Itoigawa-Shizuoka Tectonic Line and possibly by a zone of deformation in central Hokkaido. These zones of deformation are believed to represent convergence between the Eurasian (i.e., the Japan Sea and southwest Honshu Island) and North American (i.e., northwest Honshu) plates or intraplate deformation within the Eurasian plate.

Although the geologic history of southwest Japan is dominated by the accretion and uplift of deep-sea sediments, seamounts and pieces of oceanic crust, the oldest rocks are thought to be fragments of continental crust (e.g., the Hida Belt of north central Japan) (Figure 2). The Hida Belt is composed of quartzo-feldspathic gneisses and schists that record a protracted history of deformation and metamorphism. This belt at present crops out along the Japan Sea side of central Japan and is inferred to have been detached from Asia as the Sea of Japan opened in the Middle Miocene. Radiometric ages in the Hida Belt range from 400 to 150 Ma.

The continentally derived basement-like rocks characteristic of the Hida Belt are structurally underlain by a serpentinite-matrix melange (the Hida Marginal Belt; Figures 2 and 3). This belt contains blocks of various compositions and ages, although the rocks are generally no younger than Triassic (e.g., the Sangun metamorphic belt, Maizuru Belt and Akiyoshi Belt may represent relatively large blocks that correlate with the Hida Marginal Belt). Klippe and/or strike-slip slivers of the Hida Marginal Belt may also be present on central Shikoku Island as represented by the Kurosegawa Belt and possibly by the Sanbosan Belt. The Kurosegawa Belt includes Silurian to Devonian limestone and shale with faunal affinities to Australia and southern China, Silurian granites, garnet-bearing amphibolites, Carboniferous limestone, molluscan-bearing sandstone of Permian age, Permian blueschist and melange as well as shallow marine sandstone of Triassic age.

This Permian to Triassic sequence appears to be structurally underlain by a Jurassic accretionary sequence which is much more uniform in composition and metamorphic grade relative to the older sequences. The Jurassic sequence on Honshu Island is represented by the Mino and Tamba Belts and across the Inland Sea on Shikoku Island by the Chichibu and Sanbosan Belts (Figure 3). The exact correlation of these belts across the Inland Sea, however, is complicated by possibly substantial left-lateral strike-slip faulting along the Median Tectonic Line (Figures 2 and 3) in the Late Cretaceous. In any case, all of the Jurassic sequences are composed of accreted seamount complexes (pillow basalts and reef limestones) red to green ribbon radiolarian cherts and pelagic and hemipelagic shales and terrigenous trench-fill clastic deposits. They commonly show a chaotic rock facies similar to melange (Photos 2-1-1 and 2-1-2). The Jurassic sequences also contain one of the best documented stratigraphic successions of oceanic to trench sedimentation in the world as well as several beautifully exposed regional-scale duplex structures.

The relation between the Jurassic accretionary sequences and the younger Cretaceous to Tertiary sequences to the south (e.g., the Sanbagawa and Shimanto Belts) appears to be fairly complicated. Although most workers consider the contact to be a low or moderately dipping thrust fault, it is not clear which sequence is structurally highest. In fact, the structural stacking order may have been inverted as younger units were underthrust and accreted. At present, the simplest interpretation is to consider the older accretionary sequences, including the Chichibu, Kurosegawa and Sanbosan Belts, to be klippe that overlie the Cretaceous Sanbagawa and Shimanto Belts (Figure 3). Support for this interpretation comes, in part, from the structurally high position of the Jurassic accretionary sequences on

Figure 1   Computer-produced topographic relief map of the Japanese Islands. Two subduction zones, the Japan Trench and the Nankai Trough, dominate the Japanese convergent margins (provided by Japan Marine Data Center).

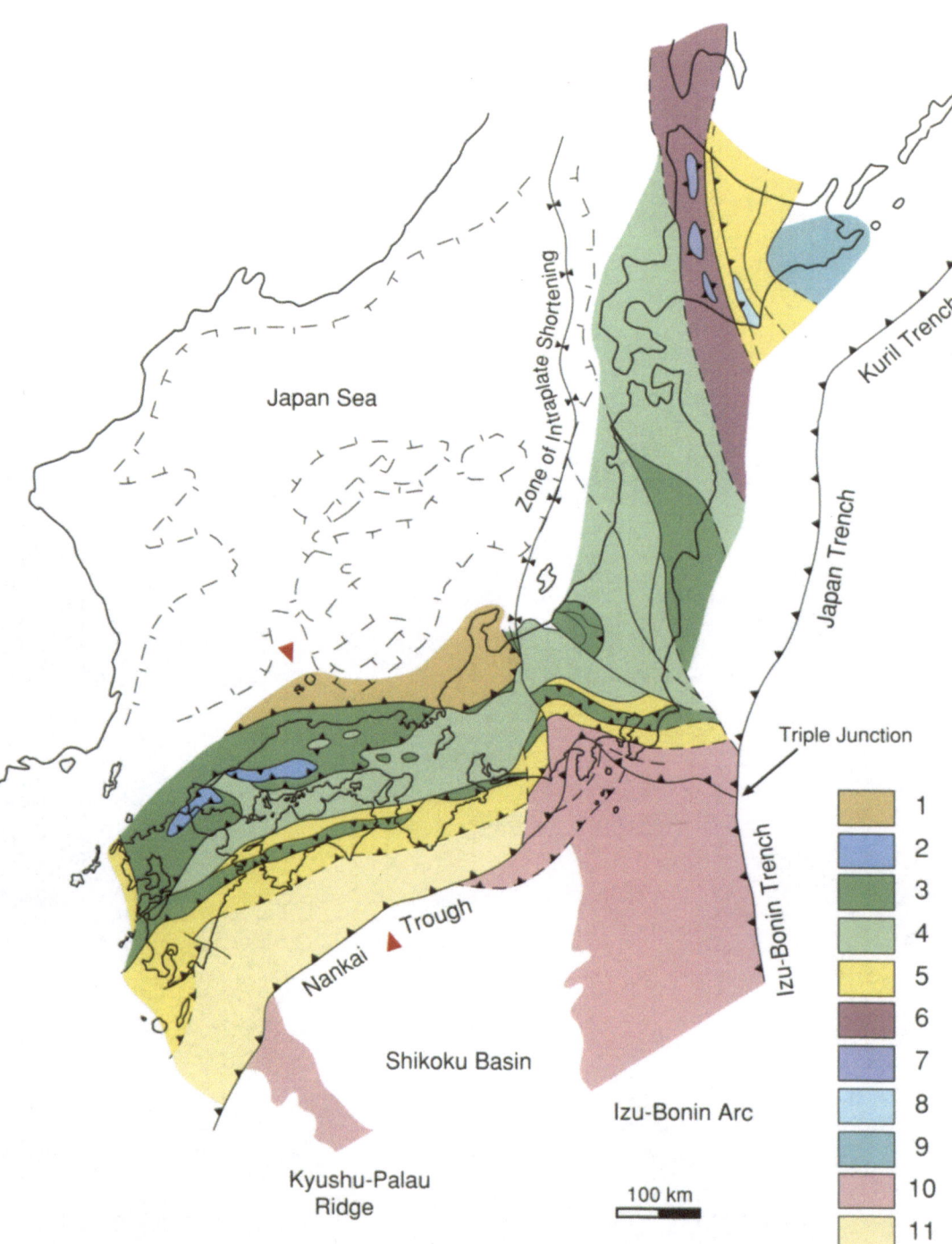

Figure 2 Simplified distribution of geotectonic units in the Japanese Islands (Taira et al., 1989). The Japanese Islands were made by successive accretion of trench sediments and oceanic plate materials. 1: Hida Belt (Paleozoic metamorphic belt and Jurassic granites), 2: Akiyoshi Belt (Paleozoic seamount and accretionary prism), 3: Sangun-Maizuru-Chichibu-Abukuma-Kitakami Belt (Paleozoic-Mesozoic accretionary prism, ophiolite and continental fragments), 4: Mino-Tanba-Ashio-Northern Kitakami-Toshima Belt (Jurassic accretionary prism), 5: Shimanto-Sanbagawa-Hidaka Belt (Cretaceous-Tertiary accretionary prism), 6: Yezo-Sorachi-Idonnappu Belt (Cretaceous ocean floor ophiolite, pelagic sequences and accretionary prisms), 7: Kamuikotan metamorphic Belt (Cretaceous ophiolitic accretionary prism), 8: Hidaka metamorphic Belt (Tertiary granulite facies metamorphic rocks), 9: Cretaceous and Paleogene clastic sequence in the Nemuro Belt, 10: Accreted arc rocks, associated sediments and volcanic arcs in the Izu collision zone, Izu-Bonin arc and Kyushu-Palau ridge, 11: Neogene accretionary prism in the Nankai Trough. Barbed broken line in the Japan Sea indicates the boundary of continental crust.

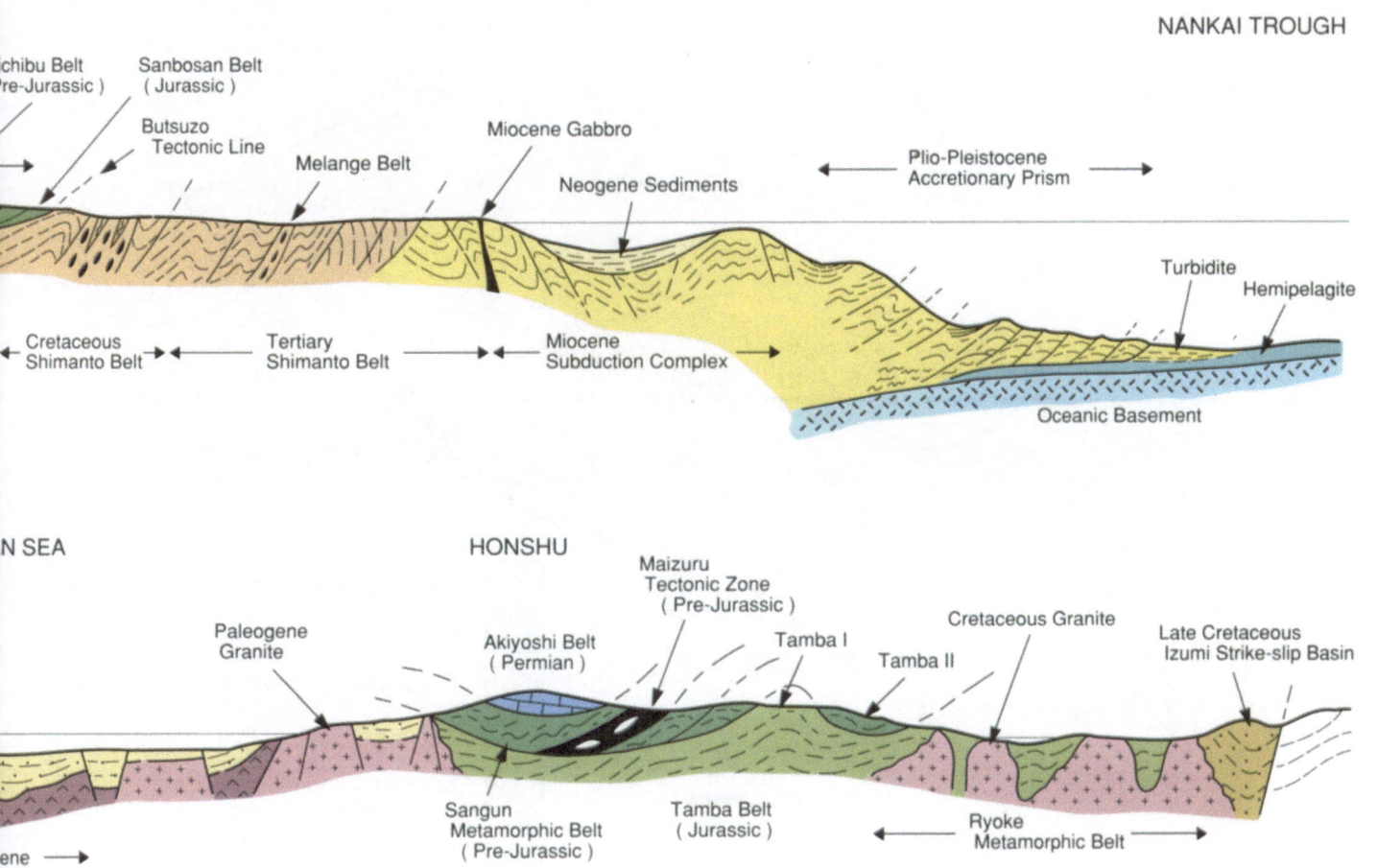

NANKAI TROUGH

ichibu Belt
Pre-Jurassic)

Sanbosan Belt
(Jurassic)

Butsuzo
Tectonic Line

Melange Belt

Miocene Gabbro

Neogene Sediments

Plio-Pleistocene
Accretionary Prism

Turbidite

Hemipelagite

Cretaceous
Shimanto Belt

Tertiary
Shimanto Belt

Miocene
Subduction Complex

Oceanic Basement

AN SEA

HONSHU

Maizuru
Tectonic Zone
(Pre-Jurassic)

Cretaceous Granite

Late Cretaceous
Izumi Strike-slip Basin

Paleogene
Granite

Akiyoshi Belt
(Permian)

Tamba I

Tamba II

Sangun
Metamorphic Belt
(Pre-Jurassic)

Tamba Belt
(Jurassic)

Ryoke
Metamorphic Belt

cene
en Tuff

Figure 3   Geological cross-sections of southwest Japan. The location of the profile is shown by an arrow in Figure 2.

Honshu Island and from an inferred gradational contact between the Sanbagawa and
Shimanto Belts on Kii Peninsula (see Figure 4).

Finally, possibly one of the most spectacular
results of the recent surge in paleontological
research in Japan is the recognition that progressively younger oceanic crust was subducted
during the Mesozoic and Cenozoic (Figure 5).
These data suggest that an active spreading
center was approaching the Japanese continental margin throughout this time. At present,
however, there is no clear or direct evidence
that this active spreading center was actually
subducted beneath Japan.

The distribution of the Upper Cretaceous
rocks in southwest Japan, including the
Shimanto Belt, is shown in the geologic map
and cross-section of Figures 2, 3 and 4. The

Upper Cretaceous rocks have been divided
into two belts: an inner belt that constitutes the
Late Cretaceous magmatic arc and an outer
belt that represents the paleo-accretionary prism
and is the main subject of this photographic
essay.

The inner belt is characterized by a wide
distribution of volcanic and plutonic rocks of
felsic composition and with radiometric ages
that are concentrated around 80 Ma; there may
also be a slight trend toward younger ages
from west to east along the belt. Rhyolitic
pyroclastic deposits associated with caldera
formation and accompanied by granodioritic
plutons are widespread in the inner belt, indicating extensive explosive volcanic activity.
The pyroclastic deposits are also sometimes
intercalated with lacustrine deposits consistent

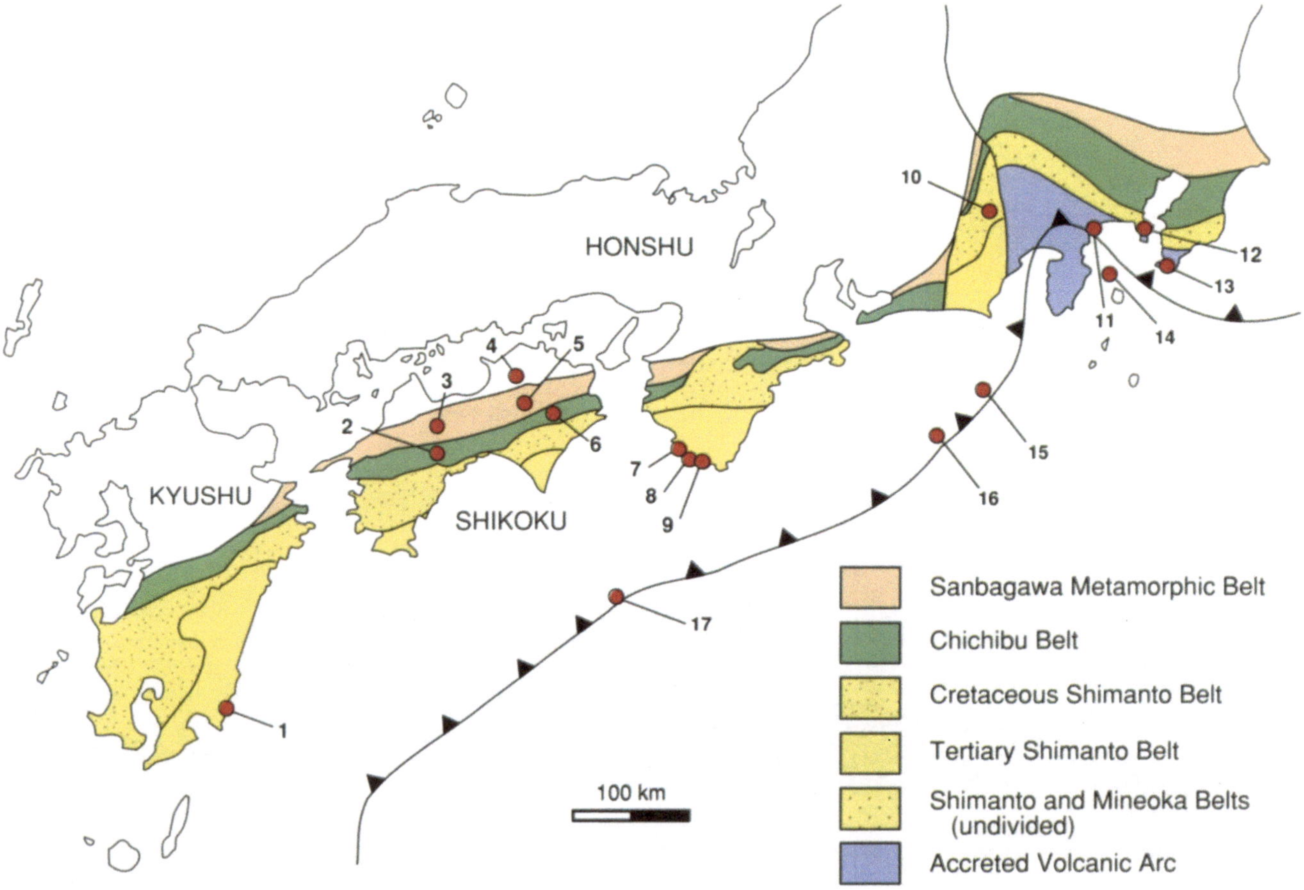

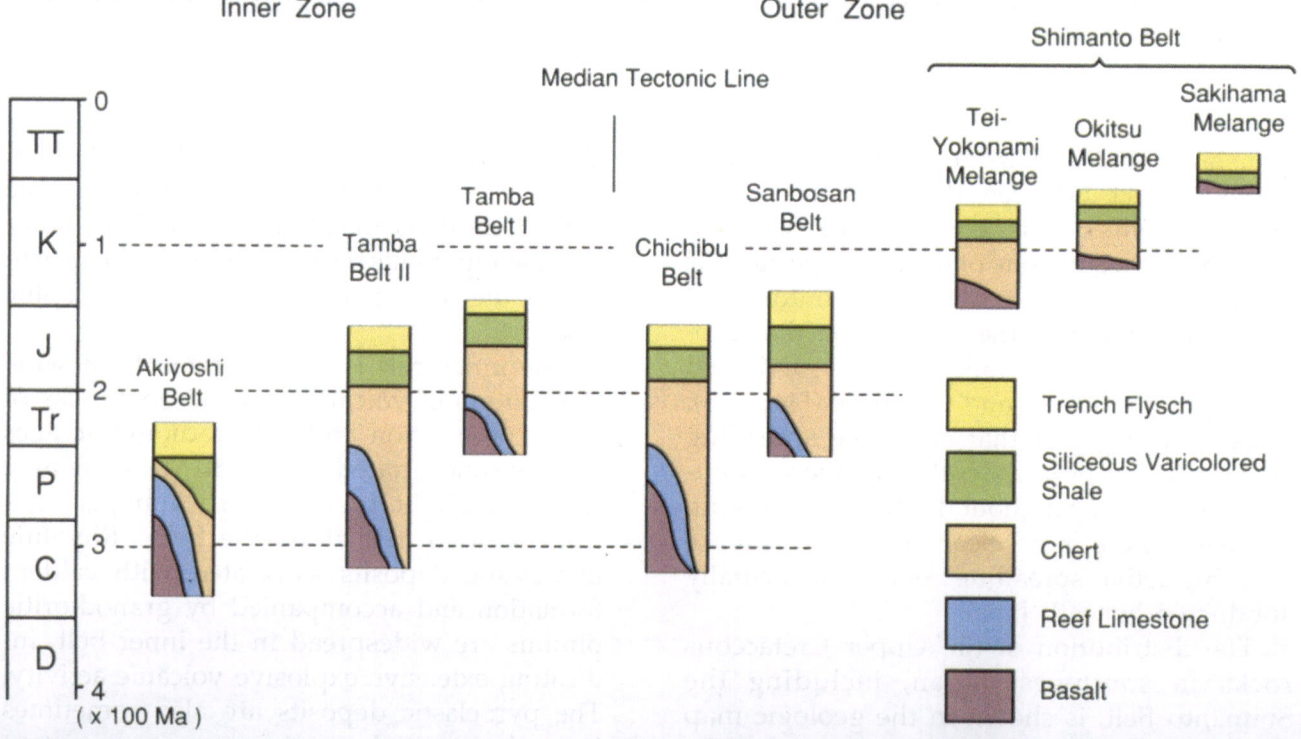

(Figure 4)

(Figure 5)

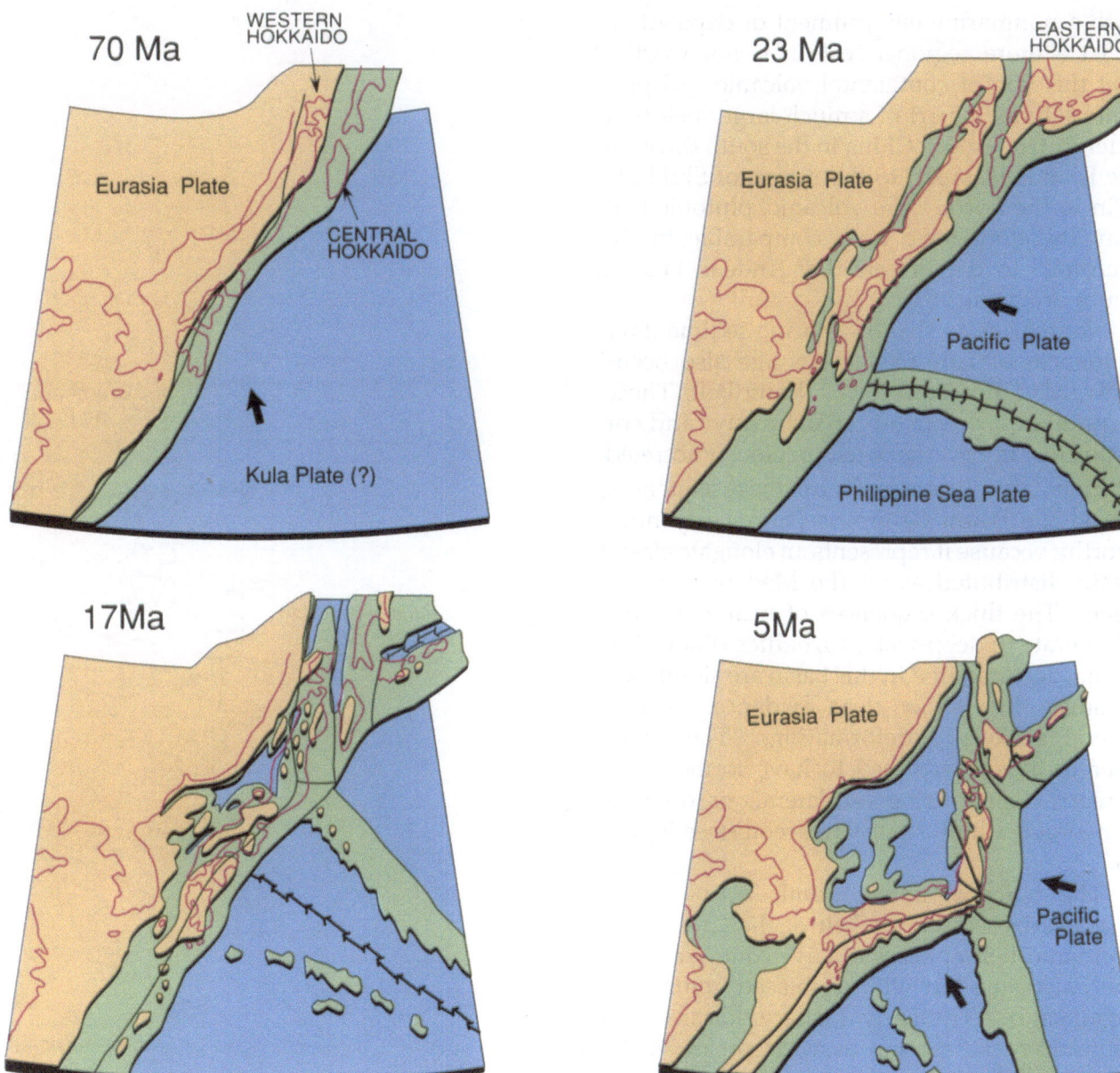

Figure 6   Tertiary development of the Japanese Island arc. (a) 70 Ma. Note the development of the Shimanto accretionary prisms during left-lateral oblique plate convergence. (b) 23 Ma. The initial rifting of the Japan Sea produced lakes west of Japan and the Izu-Ogasawara arc began migrating northward. Eastern Hokkaido had already collided against central Hokkaido at this time. (c) 17 Ma. The opening of Shikoku Basin was about end and the opening of the Japan Sea is just beginning. The opening of the Okhotsk Sea had also just started. (d) 5 Ma. Collision between the Izu-Ogasawara arc took place at this time (after Taira et al., 1989). Outline of the present configuration is drawn in red for reference.

Figure 4   Distribution of the Shimanto Belt and associated geotectonic units in southwest Japan. General locations of photos are indicated by red circles and numerals. The photo location is given in each photo caption.

Figure 5   Reconstruction of the oceanic plate stratigraphy at the southwest Japan convergent plate boundary at different time periods. The left column indicates geologic age. TT = Tertiary; D = Devonian. The horizontal axis goes generally from the northwest to the southeast across the Japanese Islands. The reconstruction was based on fossil dating of various tectonic slivers in the melange sequences.

with a nonmarine environment of deposition.

At a more regional scale it is noteworthy that this belt of continental volcanic and plutonic activity is part of a much larger belt that extends from south China in the south through the islands of Japan to the region of Shikhote Alin in the north. The volcanic/plutonic belt may therefore have been comparable in dimensions to the present-day Andean belt in South America.

Nonmarine to shallow marine sedimentary sequences of Late Cretaceous age also occur seaward of the Cretaceous volcanic belt. These sequences appear to structurally and/or depositionally overlie older, previously accreted sedimentary sequences. One of these sequences, called the Izumi Group, is particularly noteworthy because it represents an elongate clastic basin distributed along the Median Tectonic Line. The thick sequences of nearshore conglomerate to deep-water turbidites (Photo 2-1-3) that accumulated in this basin are significant because they show clear evidence for left-lateral strike-slip deformation. The basin, therefore, is interpreted to have formed as a forearc basin during left-lateral, or oblique, convergence in the Late Cretaceous (see Figure 6).

The volcanic and plutonic activity that characterizes the inner belt of the Cretaceous Shimanto Belt appears to have continued into the Paleogene but with decreased volumetric significance. The Early Tertiary volcanic and plutonic rocks are not as widespread as the equivalent Cretaceous rocks and the locus of igneous activity appears to have moved towards the present position of the Japan Sea.

The protracted history of subduction and accretion documented for the Cretaceous to Paleogene times is followed in the Middle and Late Tertiary by: (1) eastward migration of the Izu-Bonin arc as the Shikoku Basin opened in the Oligocene to Miocene; (2) clockwise and counterclockwise rotation, respectively, of the southwestern and northeastern regions of Japan during very rapid spreading in the Sea of Japan; (3) the subduction of progressively younger oceanic lithosphere, possibly associated with an active spreading center; and (4) collision of the Izu-Bonin arc with Honshu Island in central Japan (see Figure 6). It is, of

course, this last event, the collision of the Izu-Bonin arc with Japan, that dominates the present-day tectonic activity of Japan. Uplift and erosion near the collision zone (e.g., the Japanese Alps of central Honshu) is filling the Nankai Trough with sediments. These sediments are in turn being accreted to the southwest Japan forearc, forming one of the best-documented clastic-dominated accretionary prisms in the world.

N

Upper Cretaceous
Shelf-Slope Facies

Eocene
Shelf-Slope
Facies

Lower Miocene
Shelf-Slope Facies

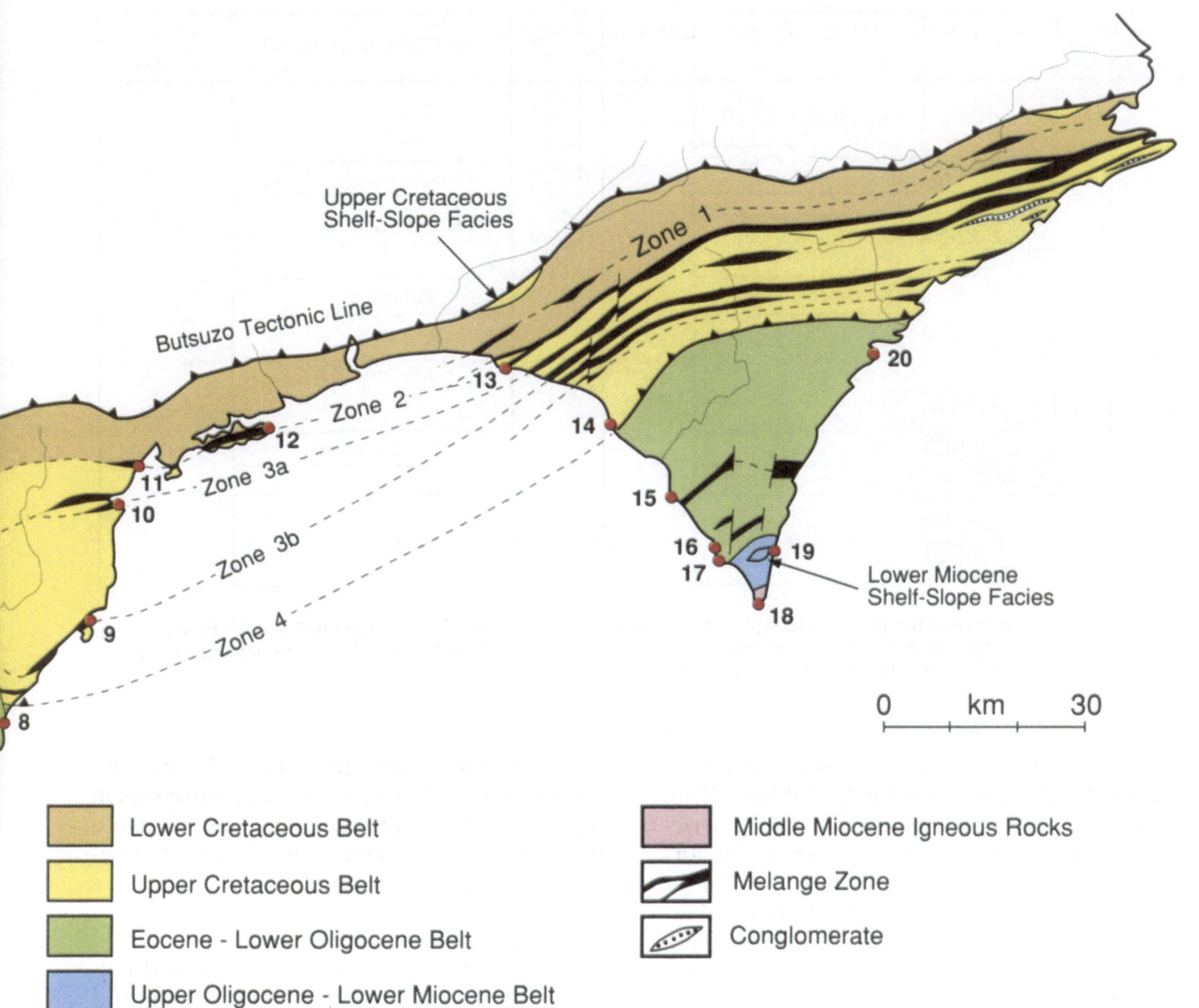

Figure 7   Distribution of the Shimanto Belt in Shikoku Island.   After Taira et al. (1988). The locations of photos are indicated by numbers and are given in each photo caption.

2-2   Geology of the Shimanto Belt

The Shimanto Belt has been subdivided into two distinct facies, based on the distribution of lithologic and biostratigraphic assemblages (Figures 3, 4 and 7). The general sedimentary and structural features of these two belts, called the Cretaceous (or northern) and Tertiary (or southern) Shimanto Belts (Figure 7), are discussed separately below.

2-2-1 The Cretaceous Shimanto Belt

The Cretaceous, or northern, Shimanto Belt has been further divided into two sub-belts, a Lower to Middle Cretaceous unit in the north and an Upper Cretaceous unit in the south (Taira et al., 1980). The northern sub-belt is composed of Neocomian to Cenomanian clastic sedimentary rocks and is characterized, in part, by the absence of associated basalt and sequences of radiolarian chert. This belt also lacks the intense deformation associated with tectonic melanges (which are common in the

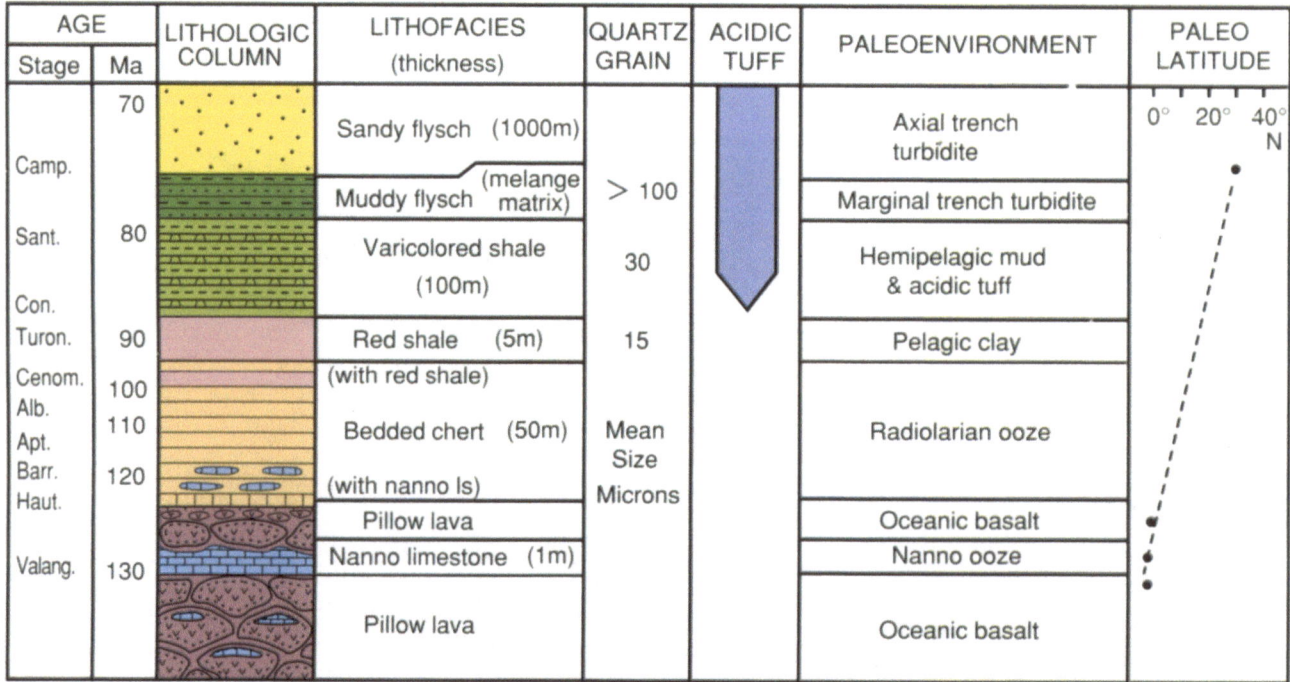

AGE		LITHOLOGIC COLUMN	LITHOFACIES (thickness)	QUARTZ GRAIN	ACIDIC TUFF	PALEOENVIRONMENT	PALEO LATITUDE
Stage	Ma						
Camp.	70		Sandy flysch (1000m)	> 100		Axial trench turbidite	0° 20° 40° N
			Muddy flysch (melange matrix)			Marginal trench turbidite	
Sant.	80		Varicolored shale (100m)	30		Hemipelagic mud & acidic tuff	
Con.							
Turon.	90		Red shale (5m)	15		Pelagic clay	
Cenom.	100		(with red shale)				
Alb.				Mean Size Microns		Radiolarian ooze	
Apt.	110		Bedded chert (50m)				
Barr.	120		(with nanno ls)				
Haut.			Pillow lava			Oceanic basalt	
Valang.	130		Nanno limestone (1m)			Nanno ooze	
			Pillow lava			Oceanic basalt	

Figure 8   Reconstructed plate stratigraphy at the time of the Late Cretaceous "Shimanto" Trench based on radiolarian ages from the melange and flysch sequences. Paleolatitude (based on paleomagnetic data) is also shown (Taira et al., 1988).

southern unit), although the clastic sequences are locally tightly to isoclinally folded. The depositional environments for the clastic sediments vary from shallow-water facies for the some of the Neocomian sediments to deep-water turbidite facies for the Cenomanian sediments. The absence of pelagic sediments (e.g., cherts) or fragments of oceanic crust (e.g., basalt) suggests that the northern belt was not deposited near an active submarine trench. Based on these observations and the regional tectonic setting we infer that this sub-belt represents a transcurrent margin environment that was deformed and accreted as subduction resumed in the latest Cretaceous.

The Upper Cretaceous Shimanto Belt also contains abundant clastic sedimentary rocks (e.g., flysch sequences) but is further characterized by the presence of several regional-scale zones of intense deformation (i.e., tectonic melanges). The melange zones typically contain fragments of ocean floor basalt, sequences of oceanic sediments, and red, green and black shale horizons. The flysch and melange zones record similar degrees of metamorphism; both are only mildly metamorphosed (typically to

zeolite facies). The age, structural style and significance of the flysch and melange zones in the southern part of the Cretaceous Shimanto Belt are presented in more detail in the following sections.

Flysch Sequences

The flysch sequences are composed dominantly of alternating beds of sandstone and shale with local horizons of conglomerate and varicolored hemipelagic shale. Fossils (radiolaria are the most common) retrieved from shale horizons interbedded with sandstone indicate ages ranging from Coniacian to Campanian with the majority of the ages being Campanian.   Trace fossils in the flysch sequences consist mostly of *Chondrites*, *Nereites* and *Helminthopsis* and are consistent with a deep-water or trench depositional environment (Katto, 1969). Paleomagnetic measurements have also revealed that the flysch units were deposited at approximately their present latitude. Finally, petrographic studies have shown that the sandstones and conglomerates are rich in rock fragments derived from acidic and intermediate volcano-plutonic terrain, probably

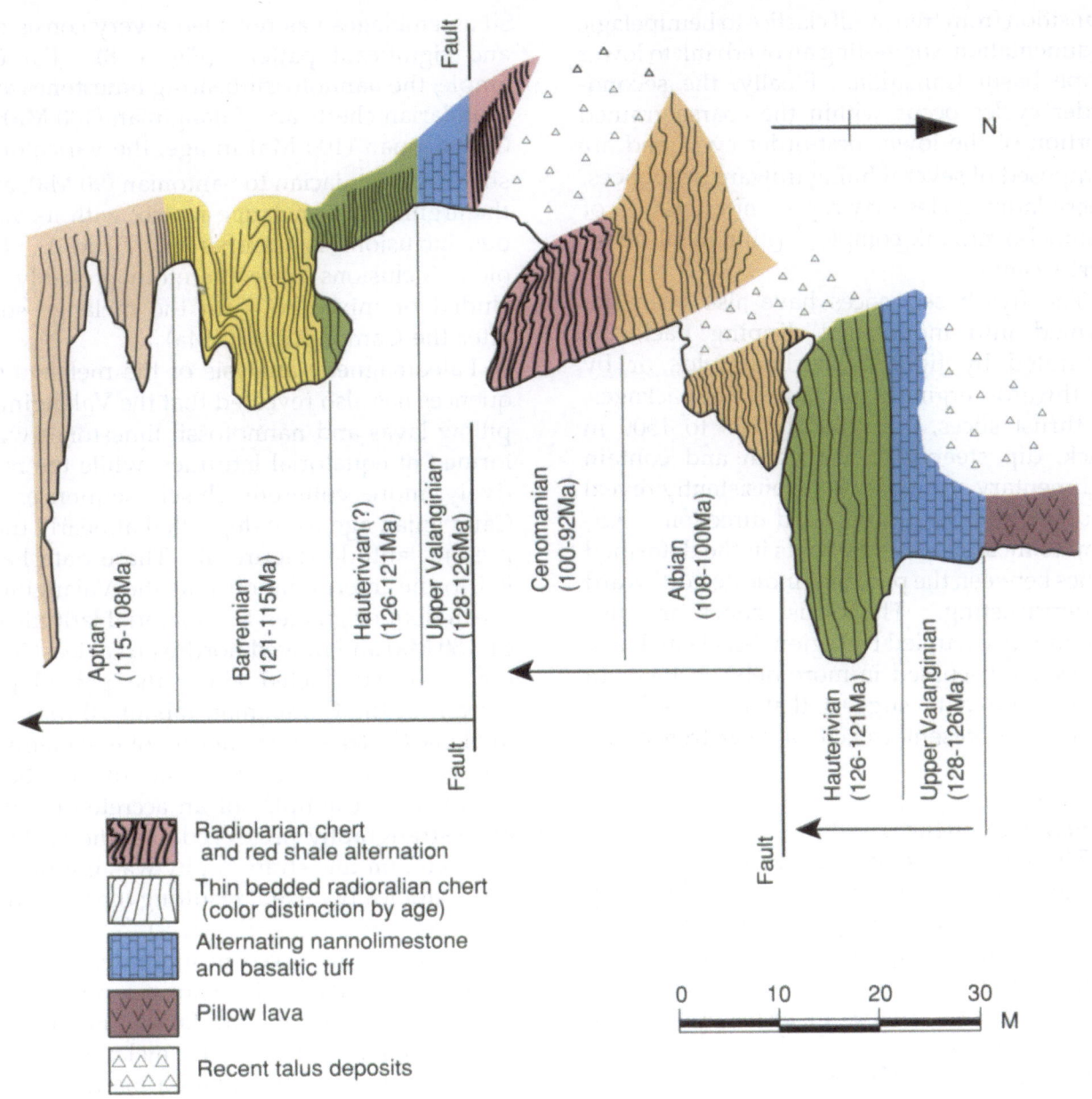

Figure 9   Radiolarian biostratigraphic age analysis of a bedded chert sequence, indicating repeated succession and southward topping (Okamura and Uto, 1982). Plan view. See Photo 5-2-2.

the Cretaceous magmatic arc complex described previously.

The varicolored hemipelagic shale layers that occur locally in the flysch sequences also contain radiolaria, but the ages can be as young as Maastrichtian. These relatively young ages and the absence of abundant clastic material in the hemipelagic sediments suggest that they represent slope sediments. Presumably, these sediments were deposited as a cover sequence on the previously deformed and accreted flysch.

Sedimentary facies analyses of the flysch sequences show two first-order vertical litho-logic cycles and several second-order cycles. The two first-order cycles are: (1) a coarsening upward cycle in the stratigraphically lower members of the Shimanto Belt and (2) a fining upward cycle in the upper members. The lower first-order cycle is interpreted to record a progressive change in the site of deposition from outer-trench margin to axial channel. The second, or upper, first-order cycle records a transition from trench-fill clastics to hemipelagic sedimentation, suggesting an overbank to lower slope basin transition. Finally, the second-order cycles occur within the coarse-grained portion of the lower first-order cycle and are composed of several fining upward sequences. These latter cycles may represent migration of channel-overbank complex within axial transport system.

The flysch sequences have also been deformed into monoclinal dipping packages separated by tight to isoclinal folds or by southward-verging thrust faults. The packages, or thrust slices, are typically 500 to 1500 m thick, dip steeply to the north and contain sedimentary structures that consistently reveal younger ages in a northward direction. Outcrop to mesoscopic scale folds in the deformed zones between the packages indicate northward underthrusting. The thrust zones are also commonly occupied by regional-scale melange zones, as discussed in more detail below. In total, these data suggest that the flysch sequences represent trench or near-trench deposits.

Melange Sequences

The melange sequences are key to understanding the origin of the Shimanto Belt; they are the *Rosetta Stone* of Japanese geology. These sequences are composed of a highly sheared argillaceous matrix in which tectonic slivers of various sizes and shapes are incorporated. The tectonic slivers are typically composed of sandstone, pillowed basalts, cherts, varicolored shales and, less commonly, limestones. Although the melange inclusions occur in a variety of shapes and sizes, their cross-sectional shapes are often asymmetric, as discussed below.

At a regional scale, the melange sequences occur as several linear belts sandwiched between sequences of flysch (Figure 7). The southern boundaries of the melange sequences are typically marked by a sharp thrust contact, whereas the northern boundary typically grades from highly sheared argillaceous rocks in the melange to more mildly deformed flysch, or "broken" formations.

Dating of the various lithologies in the Cretaceous melange with a variety of microfos-

sil assemblages has revealed a very consistent and significant pattern (Figure 8). For example, the nannofossil-bearing limestones and radiolarian cherts are Valanginian (130 Ma) to Cenomanian (100 Ma) in age, the varicolored shales are Coniacian to Santonian (90 Ma), and the argillaceous melange matrix with its various inclusions is Campanian (70 Ma). The older inclusions were therefore probably included or mixed to form the melange soon after the Campanian (70 Ma).

Paleomagnetic analysis of the melange sequences has also revealed that the Valanginian pillow lavas and nannofossil limestones were formed at equatorial latitudes, while the relatively more coherent flysch sequences of Campanian age were deposited at nearly their present latitude (Figure 8). These data have led to the interpretation that the Valanginian ocean floor originated at equatorial latitude (at ca. 130 Ma) and moved northward at least 3000 km to be subducted along the paleo-Japan margin in the Campanian (about 70 Ma). In total, the Cretaceous sequences of melange and flysch probably provide one of the best-documented examples of an accreted oceanic plate stratigraphy preserved anywhere. During accretion this stratigraphy was imbricated and tectonically mixed, resulting the formation of tectonic melanges (Figure 9).

Finally, as discussed in more detail below, the melange matrix also shows a remarkably systematic deformational fabric (Figure 10). The dominant fabrics in the melange are the lenticular-shaped blocks and a matrix characterized by a polished to scaly anastomosing foliation. Remarkably, both types of structures are often asymmetric and appear to record a sense of shear consistent with the inferred paleosubduction direction (generally northward in the Cretaceous).

2-2-2 *The Early to Middle Tertiary Shimanto Belt*

The Early Tertiary Shimanto Belt of southwest Japan is composed primarily of mildly metamorphosed (zeolite to locally higher grade), faulted and folded, but otherwise coherent sandstone-shale turbidite sequences (Taira, 1982; Taira et al., 1988). The belt crops out for over 1000 km along the southwest side

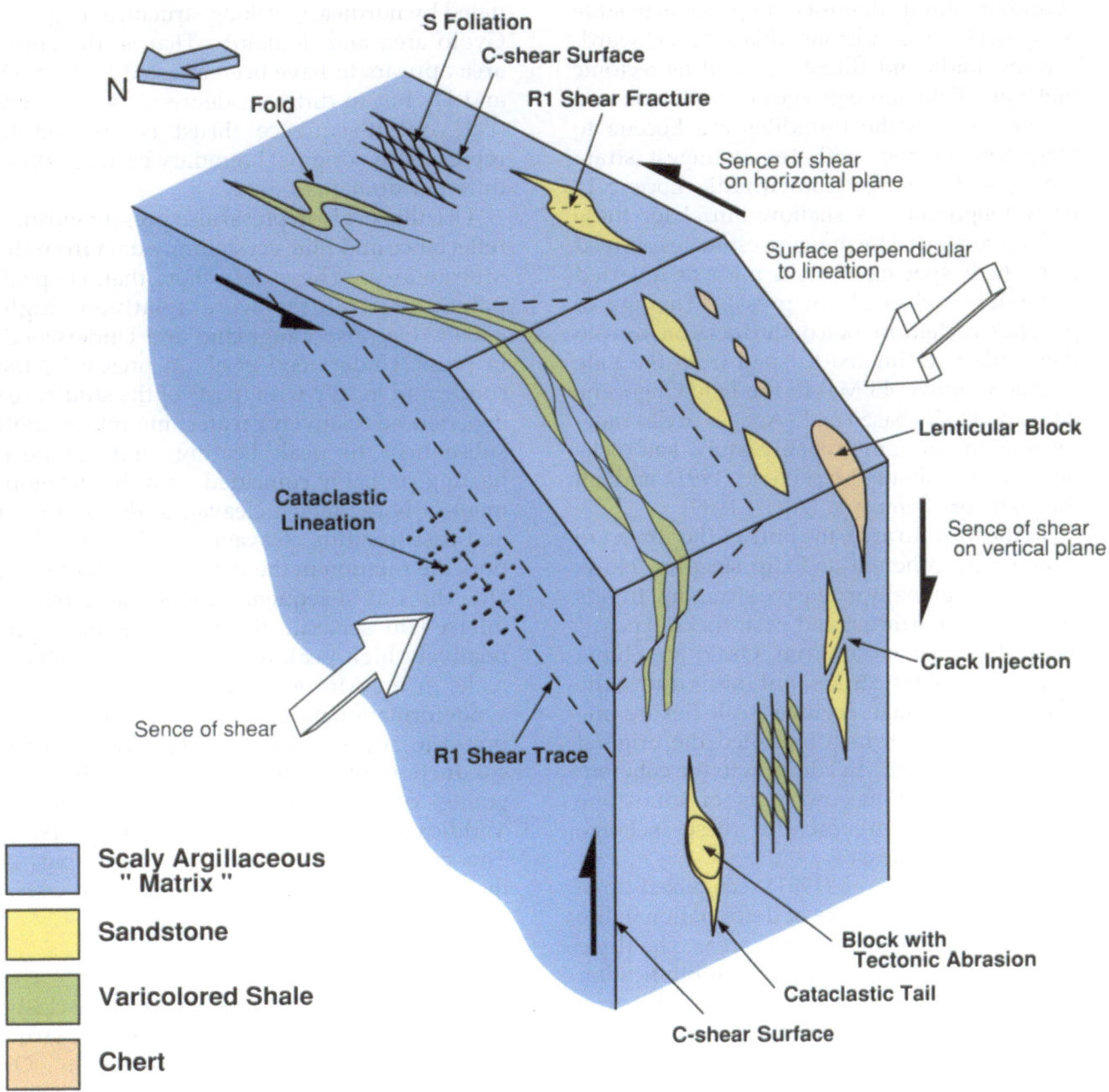

N

Fold

S Foliation

C-shear Surface

R1 Shear Fracture

Sence of shear
on horizontal plane

Surface perpendicular
to lineation

Lenticular Block

Cataclastic
Lineation

Sence of shear
on vertical plane

Crack Injection

Sence of shear

R1 Shear Trace

Scaly Argillaceous
" Matrix "

Sandstone

Varicolored Shale

Chert

Block with
Tectonic Abrasion

Cataclastic Tail

C-shear Surface

Figure 10   Mesoscale fabrics observed in the Cretaceous melange zone.

of Japan and is exceptionally well exposed in four areas: the Muroto and Ashizuri Peninsulas of Shikoku Island, the Kii Peninsula of Honshu Island, and southwest Kyushu Island (Figure 4). Although all of these areas provide abundant evidence for progressive accretion and growth of the continental margin, most of

our research, and many of the included photographs, come from the coastal exposures on Shikoku Island.

The Eocene to Oligocene Shimanto Belt of the Muroto area (Figure 7) is generally similar to the rest of the Tertiary Shimanto Belt but is somewhat unusual in that it also contains abundant slump deposits, a spectacular suite of syntectonic sandstone dikes, trenchward-verging folds and thrusts as well as tectonic melange. Paleontologic ages of radiolaria and foraminifera in the turbidites are Eocene to Oligocene in age, with the youngest strata falling in the range from Middle Eocene to Early Oligocene. A shallow burial for these rocks is suggested by pressure estimates derived from the b_0 spacing of white mica, as reported in Underwood et al. (in press). The age of penetrative deformation of the rocks on Muroto Peninsula is estimated to span from the Late Eocene (approx. 43 Ma) to the Late Oligocene (approx. 30 Ma) based on K-Ar ages of cleavage-forming micas from probably equivalent rocks on Kyushu Island (Mackenzie, 1991) and on the Ashizuri Peninsula (Agar, 1989).

In the Muroto area, the bulk of the strata are structurally coherent and dip steeply. However, two regionally extensive structural trends are present–northeast and east; rocks exposed along the east coast from Ozaki to Shiina display northwest strikes, but this is due to the effect of a late-stage, regional-scale flexure (the Muroto flexure) which modified the original strikes in this area. In addition to the coherent turbidites, three narrow belts of mudstone matrix melange are present; these melange belts strike northeast.

DiTullio and Byrne (1991) recognized three broad progressive stages of deformation in the coherent rocks of the Muroto area. The three stages are: (1) accretion-related pre-lithification imbrication with associated clastic dikes and folds; (2) major intraprism shortening expressed by regional-scale folding and pressure solution cleavage; and (3) high-angle faulting and uplift. The second stage of deformation also included the development of a major out-of-sequence thrust and an approximately 50° rotation in the shortening direction that resulted in two deformational events. The earlier, generally east-striking structures are designated as belonging to D_1, whereas the later generally northeast-striking structures are designated as belonging to D_2. The development of these D_1 and D_2 structures also appears to have been domainal, with some areas being dominated by east-striking structures (e.g. the Shiina area and domain), whereas other areas are domi-nated by northeast-striking structures (e.g., the Gyoto area and domain). That is, the entire area appears to have been affected by both D_1 and D_2, but to different degrees. A regional-scale out-of-sequence thrust is inferred to represent the original boundary between these different domains.

DiTullio et al. (in press) also present vitrinite reflectance and illite crystallinity data from the Muroto area. These data show that: (1) peak paleo-temperatures were relatively high, ~300°C (see also Laughland and Underwood, in press; Underwood et al., in press); (2) the south, and locally west, parts of the study area experienced relatively greater amounts of uplift subsequent to peak heating; and (3) peak heating probably coincided with the development of both Stage 2 cleavages (the northeast and east trending cleavages) and, by implication, the rotation in the direction of shortening and the out-of-sequence thrust fault that is inferred to separate the two domains. The relatively high peak temperatures of accreted rocks in the Muroto area and their apparent syndeformational timing led DiTullio et al. (in press) to suggest that they resulted from the subduction of relatively hot, and therefore young, oceanic crust in the Late Eocene to Middle Oligocene as suggested by Taira (1985). This crust is interpreted to have been part of the Kula or Pacific plates.

2-2-3 The Late Oligocene to Miocene Shimanto Belt

The Late Oligocene to Early Miocene clastic sedimentary rocks also form a distinct structural packet within the Shimanto Belt. On Kyushu Island and on Kii Peninsula these rocks appear to document a regionally extensive submarine sliding event (Sakai, 1987), with individual slide blocks ranging in size from outcrops to hundreds of meters in scale. On Shikoku Island, however, these submarine deposits have been overprinted by an intense tectonic deformation. Moreover, on the Muroto Peninsula, the contrast between the Eocene-Oligocene and Oligocene-Miocene is very sharp as the older rocks display a well-developed penetrative cleavage and ubiquitous veining. This contrast suggests that the older rocks originated at structurally deeper levels of the accretionary prism.

Reliable reconnaissance and detailed mapping by Taira et al. (1980), Katto and Taira (1979) and Sakai (1987) and more recently by Hibbard and Karig (1990) have shown that the Oligocene to Miocene rocks can be separated into three structurally discordant lithostratigraphic units. These include the Nabae Complex (revised and expanded from the Nabae Group of Taira et al., 1980), the Shijujiyama Formation (expanded from Katto and Taira, 1979; Sakai, 1987), and the Maruyama Intrusive Suite (new name defined in Hibbard, Karig and Taira, in prep.) (Figure 7). In addition, the Nabae Complex is divided into four informal units, including two coherent units, named the Tsuro and Misaki assemblages, and two melanges, termed the Sakamoto and Hioki melanges. The coherent units are more sandstone-rich than the mudstone-dominated melanges. Bedding is also an essential feature in the coherent units, whereas compositional layering and scaly fabric are the fundamental features of the melanges.

The age of the complex at Muroto Peninsula is loosely constrained; numerous fossils from throughout the complex indicate a Late Oligo-cene-Early Miocene age for the unit (Matsumoto and Hirata, 1972; Saito, 1980; Taira et al., 1980; Sakai, 1987; Ishikawa, 1982; Okamura and Taira, 1984). At one locale in the Sakamoto melange, a tighter age constraint is provided by Aquitanian (~23 Ma) foraminifera (Saito, 1980). In the melange units, older Eocene fossil-bearing slivers of shale are embedded in a younger late Oligocene-Early Miocene matrix, indicating lithologic mixing of two different ages.

More recent, detailed geological mapping has also revealed that there are first-order structural anomalies in the Oligocene to Miocene rocks (Hibbard and Karig, 1987, 1990). This portion of the prism is characterized by landward-verging accretion, apparent landward-verging intraprism shortening, and late phase oroclinal flexing and pervasive faulting; the oroclinal flexing was accompanied by both igneous intrusions (the Maruyama Intrusive Suite) and unusually high heat flow. The unexpected revelation of these anomalies within the prism is significant because: 1) it has direct bearing upon plate tectonic models for southwest Japan; and 2) it affords a new perspective on factors that determine the first-order structural styles in accretionary prisms.

2-1-1: View of a quarry of Permian limestone. Most of the limestone resources of southwest Japan are related to accreted seamount complexes such as this. The limestone layer is a klippe lying horizontally on the upper half of the mountain. Chichibu Belt, Mt. Torigatayama, Kochi Prefecture, Shikoku (2-Figure 4).

2-1-2: Melange rocks of the Chichibu Belt. Permian limestone tectonic slivers (A) are embedded in sheared argillaceous matrix (B) with sandstone slivers (C). Argillaceous matrix and sandstones are Jurassic trench-fill materials whereas the limestones are fragments of a seamount complex. A hammer indicates the scale. Sawadani Complex, Tokushima Prefecture, Shikoku (6-Figure 4).

2-1-3: View of a large cliff of a quarry in the Cretaceous Izumi Group. Steeply dipping and thickly bedded turbidite sandstone layers which are a part of the Late Cretaceous forearc strike-slip basin are observed. A dump truck (arrow) shows the scale. Inohana Pass, Kagawa Prefecture, Shikoku (4-Figure 4).

3 DEPOSITIONAL FEATURES IN THE TURBIDITE SEQUENCES

3-1 Sedimentary Structures Indicative of Turbidite Deposition

The coherent sequences of flysch contain a variety of sedimentary structures and sedimentary facies indicative of turbidite deposition in trench and near-trench environments. Turbidite sedimentary structures (i.e., Bouma sequences) include sole marks, graded beds and ripple laminations (Photos 3-1-1 to 3-1-3). Erosional features such as scouring, channeling and outcrop-scale cross-bedding are also common in the relatively coarse-grained beds. Finally, trace fossil assemblages include *Nereites*, *Helminthoida*, and *Paleodictyon* which are typical of a deep-water/turbiditic environment (Katto, 1969) (Photos 3-1-4 and 3-1-5).

3-2 Depositional Facies

3-2-1 *Axial Transport Facies*

The flysch sequences can be classified into the four following dominant lithologies or lithofacies:

Unit 1: Shale intercalated with thinly bedded sandstone and tuff

Unit 2: Medium bedded sandstone and interbedded shale

Unit 3: Thickly bedded sandstone and conglomerate

Unit 4: Debris flow deposits and slumps

Although the original succession of these lithofacies is complicated by later deformation, in some cases it is possible to restore the original stratigraphy. This stratigraphy appears to have been composed of first-order coarsening and a thickening upward sequence. Starting at the base, this sequence goes from: shales (Unit 1) (Photo 3-2-1), to medium bedded sandstones and shales (Unit 2) (Photos 3-2-2 and 3-2-3), to thickly bedded sandstone and conglomerate (Unit 3) (Photos 3-2-4, 3-2-7 and 3-2-8). Within Units 2 and 3 second-order fining and thinning upward depositional cycles are also common (Photos 3-2-5 and 3-2-6). The first-order cycles are similar to the coarsening upward sequence observed in the Nankai Trough, suggesting a sedimentary facies evolution related to plate convergence. This evolution is inferred to go

from an outer trench environment (Unit 1 and part of Unit 2) to a trench channel sequence (part of Unit 2 and Unit 3). The second-order cycles within Units 2 and 3 may record migration of channels within a trench floor.

3-2-2 *Transverse Transport Facies*

Debris flow deposits and olistostromes have also been recognized in the Shimanto Belt. Typically, materials in these deposits consist of abundant shale chips, sandstone blocks and reworked calcareous nodules (Photos 3-2-9 and 3-2-10). The stratigraphic association of these deposits is not certain, but it has been suggested that they were derived from the collapse of a pre-existing accretionary prism landward of the trench. Some of the olistostrome deposits contain also blocks from older accretionary prisms including Triassic limestone blocks from the Chichibu Belt. Presumably submarine canyons may have carried the debris flow deposits from higher areas up the slope to near the trench floor.

Locally, the trench sequences also contain paleocurrent indicators that record trench-perpendicular transport, and many of these show landward paleocurrent directions (Kumon et al., 1988). The exotic paleocurrents may represent turbidity currents that were reflected off the outer trench slope as the currents meandered down the axial channel.

3-3 Mineral Composition and Provenance

The mineral compositions of sandstones as well as the composition of clasts in conglomeratic horizons indicate a source dominated by both felsic volcanic-plutonic and sedimentary materials (Photo 3-3-1). There is, however, a general change in sandstone composition within the entire Shimanto Belt. For example, the Tertiary Shimanto Belt contains more sedimentary clasts in general, and more quartzose-rich sedimentary clasts in particular, than the Cretaceous Shimanto Belt (Teraoka, 1979). This change apparently documents the progressive unroofing of older accretionary complexes during the Tertiary.

3-1-1:   Turbidite Bouma sequence. The Shimanto Belt includes a variety of conglomerate-sandstone-shale interbeds, most of which have a turbidity current origin. This photograph shows a bed with a Bouma turbidite sequence from graded bedding in the lower part to current laminated in the upper part. The scale bar is 10 cm.   Upper Cretaceous to Paleogene Ohyamamisaki Formation, Aki City, Kochi Prefecture, Shikoku (14-Figure 7).

3-1-2: Flute casts on a basal surface of a turbidite bed. Such sole marks are important for identifying turbidite deposits and for determining the direction of paleocurrents. Axial, or trench-parallel paleocurrent directions dominate the flysch sequences in the Shimanto Belt but complex transverse paleocurrents, including both seaward and landward directions, have also been observed. The hammer shows the scale. Eocene Nichinan Group, Ohdotsu, Nichinan City, Miyazaki Prefecture, Kyushu (1-Figure 4).

3-1-4: Trace fossils on the basal surface of a turbidite bed which was deposited in a trench environment. *Helminthopsis* (A) to the upper left and *Paleodictyon* (B) to the lower right. Trace fossils are important indicators of depositional environments as well as stratigraphic topping direction. They also play an important role in distinguishing deep-water trench turbidite wedge depositional sites, such as this for example, from cover sequences which are deposited at much shallower water depths. Centimetric measure shows the scale. Eocene Naharigawa Formation, Kannoura, Kochi Prefecture, Shikoku (20-Figure 7).

3-1-3:   Current ripples on an upper surface of a turbidite bed. Current laminations are an important indicator for stratigraphic topping, or old → young, direction and for determining the paleocurrent direction. The hammer indicates the scale. Eocene Naharigawa Formation, Shishikui Town, Tokushima Prefecture, Shikoku (20-Figure 7).

3-1-5:   Trace fossils on the basal surface of a turbidite bed. *Nereites* feeding traces are also good indicators of stratigraphic old → young as well as deep-water depositional environments. A camera lens cap shows the scale. Eocene to Lower Oligocene Muroto Formation, Kuromi, Muroto City, Kochi Prefecture, Shikoku (16-Figure 7).

(3-2-1)

(3-2-2)

3-2-3:   A sequence of siliceous tuff beds (left side of the photo) which provides a relatively rare key bed in the Shimanto Belt. Eocene to Lower Oligocene Kurusuno Formation, Cape Zai, Tosashimizu City, Kochi Prefecture, Shikoku (7-Figure 7).

3-2-1:   Thin-bedded very fine-grained sandstone turbidite beds interbedded with shale layers. This shale-dominated facies indicates either inter-channel overbank environment or marginal trench environment. The scale bar is 10 cm. Eocene to Early Oligocene Muroto Formation, Kuromi, Muroto City, Kochi Prefecture, Shikoku (16-Figure 7).

3-2-2:   Thin- to medium-bedded coherent turbidite sequence, possibly recording a channel margin depositional environment. Note a person for scale. This sequence forms part of an antiformal stack in the Eocene to Lower Oligocene Kurusuno Formation, Cape Kanou, Tosashimizu City, Kochi Prefecture, Shikoku (2-Figure 7).

3-2-4:   Helicopter air view of vertically dipped and thickly bedded turbidite beds indicating axial trench channel depositional environments. Photo is 20 m across. Eocene Naharigawa Formation, Cape Hane, Muroto City, Kochi Prefecture, Shikoku (15-Figure 7).

3-2-5:   Finning upward second-order cycles are often observed in turbidite beds of trench channel facies. The human figure shows the scale. Eocene-Lower Oligocene Kurusuno Formation, Cape Kanou, Tosashimizu City, Kochi Prefecture, Shikoku (2-Figure 7).

3-2-6:   Transition from one finning upward cycle to the other. The uppermost part of the lower cycle is composed of thin- to medium-bedded turbidites and is overlain by thickly bedded turbidite sandstone. Each cycle is normally 10 to 50 m thick. Cape Kanou, Tosashimizu City, Kochi Prefecture, Shikoku (2-Figure 7).

3-2-7:   Vertically dipped conglomerate beds which are interbedded with very coarse-grained sandstone layers. Graded and massive conglomerate beds are occasionally intercalated with sandstone facies in the Shimanto Belt and they are good indicators of the provenance of coarse clastics. The human figure shows the scale. Upper Cretaceous to Paleogene Ohyamamisaki Formation, Aki City, Kochi Prefecture, Shikoku (14-Figure 7).

3-2-8:   A large metamorphic rock clast at the same locality as Photo 3-2-7. The metamorphic rock clasts in the Ohyamamisaki Formation is considered to be derived from the Sanbagawa Belt, placing an important constraint on the denudation of the deeper part of the Shimanto accretionary prism like the Kuma Group (Photo 7-1-3). However, the depositional age of this conglomerate beds has not been determined yet. Scale bar is 20 cm long. Aki City, Kochi Prefecture, Shikoku (14-Figure 7).

(3-2-9)

(3-2-10)

3-2-9, 3-2-10:   A conglomerate bed which includes abundant shale clast and reworked calcareous nodules. Such clast composition indicates reworking of materials from the landward slope. The matrix is sandstone. This conglomerate has been interpreted to be canyon or trench fan deposits. A camera cap shows the scale. Miocene Nabae Group, Cape Muroto, Muroto City, Kochi Prefecture, Shikoku (18-Figure 7).

3-3-1:   Photomicrograph of sandstone. Note the variety of grain composition including poly- and mono-crystalline quartz, feldspars, volcanic rock fragments, sedimentary rock fragments. The Shimanto Belt sandstones are rich in grains that were derived from both sedimentary and igneous sources. Photomicrograph is 2.5 mm across. Cretaceous Shimotsui Formation, Yokonami Peninsula, Tosa City, Kochi Prefecture, Shikoku (12-Figure 7).

4 DEFORMATIONAL FEATURES IN THE TURBIDITE SEQUENCES

The sedimentary deposits of the Shimanto Belt display a tremendous variety of deformational features, ranging from millimeter-wide cataclastically deformed stringers and tails of volcanic rocks in zones of melange to kilometer-scale slide blocks. To simplify the presentation of these structures we have organized the discussion by dividing the accompanying photographs into two broad categories that follow the division presented for the Cretaceous Shimanto Belt; that is, flysch and melange. Within these facies, however, we have also recognized subdivisions characterized by specific structures or lithologies. For example, the structures in the flysch sequences have been divided into three categories that, in part, reflect the progressive growth of convergent margins. These categories are: early, transitional and late-stage structures.

The earliest structures in submarine convergent margins typically develop in the shallowest and most seaward levels of the margin where the sediments are water-rich and barely lithified. These early structures therefore document the tectonic dewatering of the sediments. The transitional structures form after the initial stages of accretion and generally reflect deformation within the accretionary wedge. These structures typically form in sediments that are dewatered, relatively cohesive and well lithified. The late-stage structures develop as the accreted sediments are uplifted and eroded. They therefore reflect the transition from deep, high-temperature deformation processes to low-temperature, shallow level deformation processes.

Finally, we have also divided the regionally extensive and complexly deformed melange units into two groups: shale-rich melange and chert-rich melange. Such subdivision reflects, in part, the different initial stratigraphies present on the oceanic crust at the time the melanges formed. Chert-rich melange is described and discussed in a separate chapter.

4-1 Early Structures in "Wet" or Partly Lithified Sediment

The Shimanto Belt displays various deformational structures that appear to have formed in sediments that were only partly lithified at the time of deformation. Typical structures include dish-and-pillar structures, clastic dikes, thrust faults and mud-rich slump deposits.

4-1-1 Dish-and-Pillar Structures

Dish-and-pillar structures are especially well exposed on Capes Muroto and Oyamamisaki on Shikoku Island. Here, centimeter-scale dish structures are particularly well developed in medium- to coarse-grained sandstone horizons (Photos 4-1-1 and 4-1-2). The dish structures are thought to record the filling and collapse of mud/fluid-filled bubbles that formed as the coarse-grained sands were alternately saturated and drained during periods of fluid overpressuring. The pillars are inferred to be the channels that formed as the bubble collapsed. Presumably the sands were stratigraphically and tectonically overpressured very soon after deposition, and fluids escaped through an interconnected set of dishes (bubbles) and pillars. Pillar structures are in turn sometimes connected to outcrop-scale sand volcanoes (Photo 4-1-3). The dish-and-pillar structures may also grade into clastic dikes.

4-1-2 Submarine Slump Deposits

Outcrop- to regional-scale slump sequences are also associated with the turbidite sequences (Photos 4-1-4 to 4-1-6). The slump sequences can be differentiated from tectonic deformations by the chaotic character of folds in the former, the bimodal (e.g., silty sand mixed with clay) matrix of slumps, and cross-cutting relations between the slumps and structures clearly of "soft-sediment" origin (e.g., sand dikes and submarine unconformities caused by submarine channels). The regional-scale slump sequences are probably related to massive slope failure in the toe region of the accretionary prism.

4-1-3 Sandstone Dikes

Abundant sandstone dikes are also observed in some parts of turbidite sequences (Photos 4-1-7 to 4-1-13). The dikes occur in a variety of sizes and shapes and are especially common in the vicinity of thick medium- to coarse-grained sandstone horizons. Typically the dikes vary from 1 cm in width to several tens of centimeters. Individual dikes can also be traced for up to several tens of meters both parallel and oblique to bedding. In many areas the dikes appear to have been injected both up and down the stratigraphic section. Some sandstone dikes reflect the orientation of the regional stress field.

4-1-4 Outcrop-Scale Thrusting and Imbrication

Detailed field studies indicate that in some locations the early dewatering structures cross-cut zones of bedding-parallel shortening (4-1-14). These zones include outcrop-scale thrust faults and duplex structures as well as outcrop-scale folds. These observations indicate that the tectonic stresses associated with plate convergence also affected the partly lithified sediments.

4-2 Transitional Structures in Lithified Sediments

4-2-1 Regional-Scale Thrusting and Duplexing

Regional-scale structural patterns in coherent turbidite sequences bear many similarities to the structural style of foreland fold-and-thrust belts (or nonmarine accretionary wedges). One of the best-documented examples comes from the Tertiary Shimanto Belt in the Ashizuri area of Shikoku Island. Here, several distinctive tuff horizons (or "key" beds) (see Photos 3-2-2 and 3-2-3) have been correlated, and the outcrop pattern suggests a regional-scale antiformal stack or duplex (Tokunaga, in press). The antiformal stack is somewhat unusual, however, because the individual thrust packages dip seaward, as opposed to landward, suggesting a period of landward vergence in the Eocene to Oligocene.

4-2-2 Asymmetric and Sheath-Like Folds

Folds of various scales are common throughout the Shimanto Belt. The folds are typically tight to open and display an asymmetry that indicates northward underthrusting (Photos 4-2-1 to 4-2-9). Sheath folds have also recently been documented, although they are generally relatively rare in the Shimanto

Belt (Hibbard and Karig, 1987) (Photo 4-2-10). Several particularly well-exposed examples are present in the Late Oligocene to Early Miocene Shimanto Belt at Cape Muroto. Like the antiformal stack discussed above, the sheath-like folds at Muroto are also unusual because they record landward vergence. Asymmetric folds and sheath-like folds are sometimes associated with outcrop-scale layer-parallel shortening (Photo 4-2-11).

4-2-3 Pressure-Solution Cleavage

The tectonic folds are also often associated with the development of a spaced to slaty-like cleavage (Photo 4-2-12). The cleavage is best developed in shale-rich layers and, in thin section, is defined by the preferred orientation of phyllosilicates and insoluble residue material. Phyllosilicate fibers are also locally developed around diagenetic framboidal pyrite (Photo 4-2-13). In the area of Cape Ashizuri, regional folding predates the development of the regional cleavage, and fold axes and cleavage trajectory are not concordant. The trend of the cleavage in the Muroto area also appears to have rotated about 40°, perhaps as a result of a change in plate motions in the Middle Tertiary.

4-3 Stratal Disruption, Broken Formations and Shale-Rich Melanges

Zones of stratal disruption, broken formation and melange are characterized by a suite of gradational structures, ranging from slightly disrupted beds that may be continuous for dozens of meters to highly broken and discontinuous units (Photos 4-3-1 to 4-3-3). In the more deformed varieties, more competent layers are often necked, or boudined, and individual boudins can be rotated relative to the initial sedimentary bedding fabric (Photos 4-3-4 and 4-3-5). In some cases, necking and folding appear to be contemporaneous and fold hinges are isolated from any initial layering. The fold hinges appear to "float" in a more ductilely deforming matrix of shale (Photos 4-3-6 to 4-3-8).

Brecciation of blocks of various sizes is also commonly observed. At the outcrop-scale, sandstone blocks show a zig-zag pattern of crack filling and fragmentation (Photos 4-3-9 and 4-3-10).

At a microscopic scale the disruption appears to have been accommodated by both cataclasis and particulate flow (Photos 4-3-11 to 4-3-13). Individual quartz and feldspar grains are often fractured or undulatory, and fractures can be filled with shale from the matrix, attesting to the mobility of the shale matrix, or the fractures may be filled with newly crystallized minerals (e.g., calcite or quartz). The scaly fabrics in thin section are also distinctive. The phyllosilicates that define these fabrics are typically not strongly foliated or aligned; instead, they show a flow-like texture with many micro-scale folds and "swirls" apparent. These textures and fabrics suggest that the sediments were highly overpressured and ductile when the shale-rich melanges and scaly fabrics formed.

The compositions of the zones of stratal disruption and broken formations are also distinct in comparison with the melanges recognized in the Mesozoic Shimanto Belt. For example, the Tertiary melange belts are gener-

ally dominated by thick zones of sheared, or scaly, shale, whereas the Cretaceous melanges (see below) also contain abundant chert, basalt and varicolored shale.

Finally, we should emphasize that, although the Tertiary and Cretaceous melange zones contain different lithologies, these zones probably developed through similar processes. Specifically, we envision both melanges forming relatively early in the accretionary history with tectonic mixing processes (e.g., faulting and penetrative shearing) being more dominant than sedimentary processes (e.g., gravity sliding and slumping). We also consider that many of the melanges more formed at or near the base of the accretionary prism prior to complete lithification of the sediments (e.g., Fisher and Byrne, 1987). Thus, we consider the melanges to be tectonic shear zones that formed in partially lithified sediments. In this interpretation the different lithologies of the Tertiary and Cretaceous melanges simply reflect the different oceanic stratigraphies at the time of subduction.

4-1-1:
Dewatering pipelets, pillars, in a thick bedded, coarse-grained sandstone layer which is composed of several successive turbidity current depositions (amalgamated beds). Note that the pipes are truncated by an upper sandstone layer. The pipes show at least two generations; one is vertical (A) and the other is creeping (B). The vertical pipelets crosscut the creeping ones. The slight change in color within the pipelets reflects the removal of fine-grained clay particles during fluid flow. Scale bar is 15 cm long. Eocene-Lower Oligocene Muroto Formation, Cape Gyodo, Muroto City, Kochi Prefecture, Shikoku (17-Figure 7).

4-1-2: Dewatering pipes in a thick-bedded, coarse-grained sandstone layer. Note that small pipes converge into a larger pipe, suggesting a flow accumulation phenomenon. A coin (2.5 cm diameter) shows the scale. Eocene-Lower Oligocene Muroto Formation, Cape Gyodo, Muroto City, Kochi Prefecture, Shikoku (17-Figure 7).

4-1-3: A view of the upper surface of a thick, coarse-grained sandstone layer. Light-colored irregular patterns are a plan view of dewatering structures which appear as pipes or pillars in vertical cross-section. Some parts of these dewatering structures resemble small sand volcanoes (A). Each color unit on the scale bar is 10 cm. Uppermost Cretaceous-Paleogene Ohyamamisaki Formation, Aki City, Kochi Prefecture, Shikoku (14-Figure 7).

(4-1-5)

(4-1-6)

(4-1-4)

4-1-4:   Slump beds in the Eocene-Lower Oligocene Muroto Formation. These slump beds consist exclusively of sandstones and shales or their intermixtures and occur within a coherent turbidite facies, indicating a massive failure of the trench sequence at the toe of the accretionary prism. No exotic clasts are found. Scale bar is 1 m long.   Kuromi, Muroto City, Kochi Prefecture, Shikoku (16-Figure 7).

4-1-5:   Chaotic dispersion of rootless fold hinges and pinched sandstone layers in a muddy matrix which is a bimodal mixture of muds and sands. Such lithologies are diagnostic features of slump-debris flow beds. 15-cm scale bar. Eocene-Lower Oligocene Muroto Formation, Kuromi, Muroto City, Kochi Prefecture, Shikoku (16-Figure 7).

4-1-6:   Photograph of rootless and refolded fold hinge in a slump fold. Note also the pinch and swell structure of sandstone layers.   Complex refolding is quite common in a slump bed, which can also be used to distinguish between tectonic folding and slump folding. A hammer shows the scale. Eocene-Lower Oligocene Muroto Formation, Kuromi, Muroto City, Kochi Prefecture, Shikoku (16-Figure 7).

(4-1-8)

(4-1-9)

(4-1-7)

4-1-7: Bedded sandstone turbidite deposits intruded by a clastic dike at Cape Gyodo, Muroto Peninsula, Shikoku Island. The dike is about 20 cm thick and extends for tens of meters along the strike. The Cape Gyodo area exposes literally hundreds of clastic dikes and sills, and an equal number appear to have been injected up and down the stratigraphic section. The dike swarm in Cape Gyodo has been interpreted to be generated under a subhorizontal principal compressional stress axis. Eocene-Lower Oligocene Muroto Formation, Cape Gyodo, Muroto City, Kochi Prefecture, Shikoku (17-Figure 7).

4-1-8: Complex network of sandstone sills. Note shale xenolith (A) within a dike "pod". A trace of bedding can be seen next to the hammer handle and is approximately parallel to the sand intrusions. Eocene-Lower Oligocene Muroto Formation, Cape Hane, Muroto City, Kochi Prefecture, Shikoku (15-Figure 7).

4-1-9: Intrusion behavior of a sandstone dike. Note a change from subvertical to horizontal intrusion. Layer parallel sandstone sills are sometimes mistaken as massive beds. 15-cm scale bar. Eocene-Lower Oligocene Muroto Formation, Cape Gyodo, Muroto City, Kochi Prefecture, Shikoku (17-Figure 7).

4-1-10: Aerial photograph of an extensive slump deposit in the Eocene Shimanto Belt, which is intruded by a 20-cm-wide clastic dike. The dike extends for tens of meters at this exposure. Eocene-Lower Oligocene Muroto Formation, Kuromi, Muroto City, Kochi Prefecture, Shikoku (16-Figure 7).

4-1-11: Outcrop-scale fold hinge in the Eocene-Lower Oligocene Muro Group on Kii Peninsula. Note also the presence of 6 or 7 small dikelets that were injected upward (regional bedding in this area is overturned) from the sandstone (20 cm thick) in the middle of the photograph. The intersection of the dikes with bedding is parallel to the fold axis defined by the bedding, suggesting a seemingly genetic relation between diking and folding. The dikelets, however, have been sheared in opposite directions along the two fold limbs, and therefore are interpreted to predate folding and are only coincidentally parallel to the regional-scale fold axes. This fold appears in Photo 4-2-2. Amadori, Susami Town, Wakayama Prefecture (8-Figure 4).

4-1-12: A large sandstone dike intruding into a coherent turbidite sequence. Note that the dike is necked due to later compaction and that locally fractures are filled by quartz veins. One-meter-long scale bar. Eocene-Lower Oligocene Muroto Formation, Cape Gyodo, Muroto City, Kochi Prefecture, Shikoku (17-Figure 7).

4-1-13:
Photograph of a sandstone dike that cuts bedding at a high angle. The dike has clearly been intensely folded after it was intruded. Folding was caused by both perpendicular bedding compaction and shortening related to cleavage formation. Cleavage is nearly parallel to bedding at this locality. Scale bar is 10 cm long. Eocene-Lower Oligocene Muroto Formation, Cape Hane, Muroto City, Kochi Prefecture, Shikoku (15-Figure 7).

4-1-14:
Injection of a sandstone dike into deformed rocks. Sandstone dikes in the Muroto Formation intrude into slump deposits and some of the early tectonic folds. A meter-long scale bar. Eocene-Lower Oligocene Muroto Formation, Cape Gyodo, Muroto City, Kochi Prefecture, Shikoku (17-Figure 7).

4-2-1:   A regional-scale asymmetric fold that occurs on the overturned limb of a higher-order overturned syncline. The higher-order synclinal axis is to the left (or north). The strata involved are about 15 m thick and occur in the Eocene-Lower Oligocene Muro Group, Amadori, Susami Town, Wakayama Prefecture, Kii Peninsula (8-Figure 4).

4-2-2:   Early tectonic fold at the same location as 4-2-1. A hammer (A) shows the scale. Arrow B indicates the hold hinge shown in Photo 4-1-11. Eocene-Lower Oligocene Muro Group, Amadori, Susami Town, Wakayama Prefecture, Kii Peninsula (8-Figure 4).

(4-2-3)

(4-2-4)

(4-2-5)

4-2-3, 4-2-4, 4-2-5: Early tectonic fold. Note the change in bed thickness, indicating ductility of beds. The hammer shows the scale. Eocene-Lower Oligocene Muro Group, Amadori, Susami Town, Wakayama Prefecture, Kii Peninsula (8-Figure 4).

4-2-6: Early small-scale asymmetric tectonic fold within steeply bedded turbidites. The human figure shows the scale. Eocene Naharigawa Formation, Kannoura, Kochi Prefecture, Shikoku (20-Figure 7).

4-2-7:   A kite-flown photograph of early tectonic folds exposed on the surface of a marine terrace. From Kodama et al. (1988). Eocene-Lower Oligocene Muroto Formation, Cape Gyodo, Muroto City, Kochi Prefecture, Shikoku (17-Figure 7).

(4-2-8)

(4-2-9)

4-2-8, 4-2-9: Close-up view of the early tectonic folds shown in Photo 4-2-7. Note the progressive change in direction of fold axis, indicating the sheath-like nature of the fold. The scale is shown either by a 1-m-long bar or a hammer. Eocene-Lower Oligocene Muroto Formation, Cape Gyodo, Muroto City, Kochi Prefecture, Shikoku (17-Figure 7).

4-2-10: Plan view of cross-section of a sheath fold. Note that the upper (this is actually the lower part of sheath) portion of the sheath is truncated by the fault (A). Sheath folding is a useful indicator of tectonic vergence. Miocene Nabae Group, Cape Muroto, Muroto City, Kochi Prefecture, Shikoku (18-Figure 7).

46

4-2-11:
View of an upper surface of a thin turbidite bed showing deformed linguoid ripples. Note slicken lines on quartz veins (A), indicating layer parallel shortening, as shown by arrows. Eocene-Upper Oligocene Muroto Formation, Cape Gyodo, Muroto City, Kochi Prefecture, Shikoku (17-Figure 7).

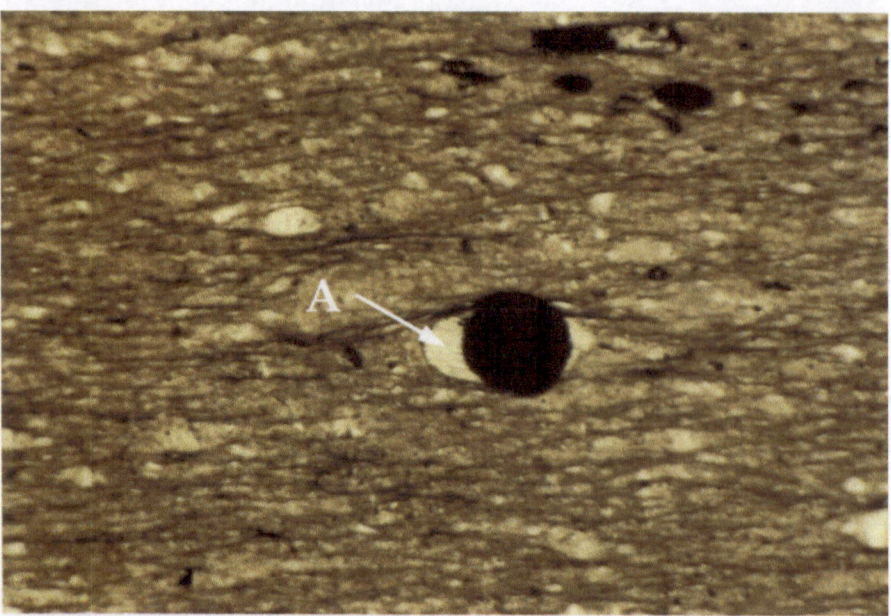

4-2-13: Photomicrograph of solution cleavage. Note phyllosilicate fibers (A) around the pyrite grain. Eocene-Upper Oligocene Muroto Formation, Cape Gyodo, Muroto City, Kochi Prefecture, Shikoku (17-Figure 7).

4-2-12: Slaty cleavage developed in shales. Cleavage dips moderately to the right. Eocene-Upper Oligocene Muroto Formation, Cape Gyodo, Muroto City, Kochi Prefecture, Shikoku (17-Figure 7).

4-3-1: Necking and block rotation due to normal faulting. Layer parallel flattening is very common in the Shimanto Belt. A hammer shows the scale. Uppermost Cretaceous to Paleogene Ohyamamisaki Formation, Aki City, Kochi Prefecture, Shikoku (14-Figure 7).

48

4-3-2: Necking and vertical fractures indicating layer-parallel flattening. Scale bar is 1 m long. Uppermost Cretaceous to Paleogene Ohyamamisaki Formation, Aki City, Kochi Prefecture, Shikoku (14-Figure 7).

4-3-3: Folded layer-parallel flattened strata. Layer-parallel flattening often predates later deformation. Eocene Tanokuchi Formation, Shirahama, Saga Town, Kochi Prefecture, Shikoku (14-Figure 7).

4-3-4: Complex fold pattern and stratal mixing in the shale-rich part of a broken formation. A hammer shows the scale. Eocene Tanokuchi Formation, Shirahama, Saga Town, Kochi Prefecture, Shikoku (8-Figure 7).

4-3-5: Lens-shaped sandstone blocks in argillaceous matrix. These lenses are aligned semi-continuously, suggesting that once they composed a single layer. A human figure shows the scale. Eocene Tanokuchi Formation, Shirahama, Saga Town, Kochi Prefecture, Shikoku (8-Figure 7).

4-3-6: Rootless and dislocated fold hinges indicating continuous ductile deformation. A hammer shows the scale. Upper Cretaceous Okitsu Melange, Okitsu, Kubokawa Town, Kochi Prefecture, Shikoku (9-Figure 7).

4-3-7: Isolated fold hinge which shows brecciation and necking. Fractures are filled by the shale in the surrounding matrix. The hammer shows the scale. Eocene Tanokuchi Formation, Shirahama, Saga Town, Kochi Prefecture, Shikoku (8-Figure 7).

4-3-8:   Isolated fold lens. Such a lens can occur on various scales, from microscopic to tens of meters long. A human figure shows the scale. Eocene Tanokuchi Formation, Sawano-tohge, Ohgata Town, Kochi Prefecture, Shikoku (15 km north of 7-Figure 7).

4-3-9:   A block of sandstone showing an initial stage of brecciation. Note the development of cracks (A) which are filled by argillaceous matrix. A camera lens cap shows the scale. Eocene Tanokuchi Formation, Shirahama, Saga Town, Kochi Prefecture, Shikoku (8-Figure 7).

4-3-10: Brecciation of a sandstone block. Note the network of fractures which are filled by shale matrix, possibly indicating deformation under high pore-fluid pressure. Scale bar is 15 cm long. Eocene Tanokuchi Formation, Shirahama, Saga Town, Kochi Prefecture, Shikoku (8-Figure 7).

4-3-11: Open-nicol photomicrograph showing sandstone brecciation. Note that the larger sandstone body is highly fractured and that the sand appears to grade into the smaller blocks of sand that are isolated in the mud matrix. Also note that there are no strong linear or planar fabrics. Scale bar is 1.0 mm long. Eocene Tanokuchi Formation, Shirahama, Saga Town, Kochi Prefecture, Shikoku (8-Figure 7).

4-3-12: Open-nicol photomicrograph of matrix of shale-rich melange. Note the anastomosing network of the fractures and the random orientation of the long axis of silty grains. Scale bar is 2.5 mm long. Eocene Tanokuchi Formation, Shirahama, Saga Town, Kochi Prefecture, Shikoku (8-Figure 7).

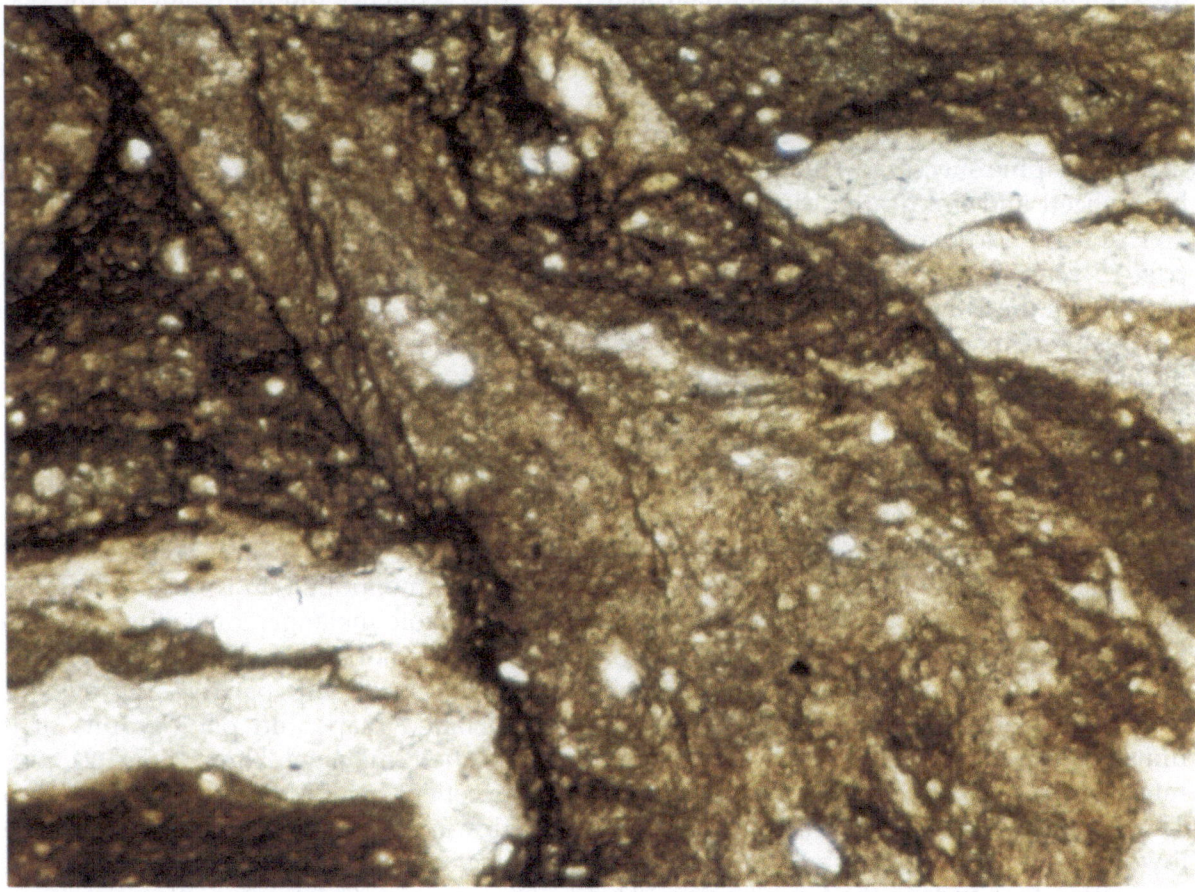

4-3-13: Open-nicol photomicrograph of infilling of fractures in the matrix of a shale-rich melange. Note the offset of quartz veins and shale infilling within the fracture. Scale bar is 2.5 mm long. Eocene Tanokuchi Formation, Shirahama, Saga Town, Kochi Prefecture, Shikoku (8-Figure 7).

5 FORMATION OF CHERT-RICH MELANGE SEQUENCES

Zones of concentrated deformation that we have recognized as chert-rich melange form one of the significant components of the accretionary prism in southwest Japan. Besides being zones of concentrated deformation, the chert-rich melanges contain a distinctive suite of lithologies that appear to record the initial stratigraphy of the subducting oceanic lithosphere. The distinctive structural and stratigraphic aspects of these units are discussed separately below and shown as examples in the photographs.

5-1 Melange Lithologies and Oceanic Plate Stratigraphy

At outcrop- and thin-section scales the Mesozoic chert-rich melanges display a "block-in-matrix" texture that is typical of many melange belts around the world. The blocks, or tectonic slivers, include a variety of rock types that occur as elongate lensoids in a well-foliated, planer matrix of black, red or green shale. The blocks in the melange can include varicolored shale, radiolarian chert, red shale, pillowed basaltic lava and nannoplankton limestone. The blocks, however, do not occur in stratigraphic succession, but appear to have been sliced and imbricated during accretion.

Extensive radiolarian age dating over the past 15 years (with over 20,000 samples processed from the Shimanto Belt alone), however, has resulted in a remarkably consistent age-lithology relation (Taira et al., 1980). A summary of these age relations and the typical outcrop appearance of the different units is presented in figures (Figure 9; Taira et al., 1988) and photographs (Photos 5-1-1 to 5-1-8).

On a more regional-scale the Mesozoic melange units show an age-lithology relation that changes from north to south across the Shimanto Belt (Taira et al., 1988). Four different melange units have been recognized (Figure 7):

Zone 1: This is the northernmost melange unit and contains the oldest oceanic slivers interbedded with Tithonian red pelagic shale. The deposition of the red shales is then followed by a long history of radiolaria-rich chert deposition that ends in the Cenomanian. This is followed by Turonian red pelagic shale and Coniacian to Santonian hemipelagic sediments.

Zone 2: This zone is probably the best-studied melange unit in the Shimanto Belt because the radiolarians are very well preserved (Figure 8). The basaltic rocks in this zone appear to be Valanginian and are overlain by nannoplankton limestone, basaltic tuff, limestone and chert interlayers that grade upward into Turonian red shale and acidic tuff which are in turn followed by deposition of Campanian shale and fine-grained turbidites.

Zone 3: The oldest rocks found in this zone are Albian to Cenomanian cherts that are overlain by red shales. The black shale matrix of this zone is Campanian, suggesting a much narrower age range for the oceanic stratigraphy.

Zone 4: This is the southernmost zone of the Cretaceous Shimanto Belt, and the oldest tectonic slivers are probably Cenomanian to Turonian. The overall amount of chert in this zone is relatively small, and basaltic tuff and red shale are more common, suggesting that this sequence did not reside on the ocean floor for an extended period of time. This is supported by the Campanian age of the shale matrix.

In the Tertiary melange belt, no pelagic chert has been found in Shikoku. The age difference between red tuffaceous shale associated with pillow basalt and the black shales of the melange matrix is also very small. As a result, it has been suggested that the subducting oceanic lithosphere was very young and active ridge subduction might have occurred (Taira, 1985).

5-2 Imbrication of the Oceanic Plate Stratigraphy

As mentioned above, the lithologies of the Mesozoic melange zones appear to occur in a rather random order; however, faults and shear surfaces are also pervasive at every outcrop. Detailed maps and photographs of the best-studied melange zone (Zone 2) provide an excellent example of the structural complexities in these zones (Photo 5-2-1). The typical occurrence of variously sized tectonic slivers of oceanic material within the melange zones is well illustrated, and a chert sequence within this zone appears to be up to 250 m thick

(Photo 5-2-2). Detailed radiolarian biostratigraphy in the chert sequence, however, suggests that it is tectonically repeated several times (Figure 9), and the original thickness of the layer was probably no more than 70 m. This particular chert sequence also shows a younger age in the southward direction. Other chert beds and pillow lavas distributed in the northern half of this melange show a younger age in the northward direction. Thus, it is suggested that the melange zone was first imbricated and then folded to form a regional-scale anticline. The hinge zone is poorly exposed and probably completely sheared out.

At an outcrop scale, the internal structure and contacts between the different rock types are much more complicated (Photos 5-2-3 to 5-2-9). Intermixing of the different rock types is also very common. For example, small lenses and blocks of cherts and basalts are incorporated into the matrix as "elongated fish" and abraded "balls" that are often associated with cataclastically produced powders. The various rock types can also be imbricated at any scale and the matrix can be mobilized or remobilized to form locally derived shale intrusions. Such intermixing and cataclastic grinding, which are sometimes misinterpreted as original sedimentary or volcanic interlayering, are a very common, if not integral, part of the melange zone. The clear age differences between the rock types observed in the melange zone, however, exclude the possibility that this complicated mixing resulted from sedimentary and volcano-plutonic processes.

Although the melange zones appear to record a chaotic or random mixture of rock types, two remarkable facts indicate that the melange was formed by an orderly process. First, the melange zones are characterized by a remarkably uniform suite of rock assemblages, as described above, that occur in tectono-stratigraphic horizons that are traceable laterally for a distance of up to 250 km. Second, the outcrop- to microscopic-scale structures observed in the melange show a striking uniformity, as discussed below.

5-3 Melange Fabrics

The three-dimensional outcrop-scale fabrics in the melange often show consistent asymmetric patterns. This was first pointed out by Taira et al. (1988) in a reconnaissance study and later more extensively studied by other workers (e.g., Kimura and Mukai, 1991; Kano et al., 1991). Fisher and Byrne (1987) and Byrne and Fisher (1990) have also documented similar asymmetric structures in tectonic melanges in Alaska. The asymmetric fabrics occur as two types that in very general terms are geometrically similar to asymmetric fabrics recognized in mylonites (termed S-C fabrics) and in brittle shear zones (R_1 and R_2 shears). Typical examples of these fabrics can be seen in outcrops, hand specimens and thin sections. The main features of the mesoscopic scale fabrics are summarized in Figure 10 and in the associated photographs (Photos 5-3-1 to 5-3-19). Regional mapping of the asymmetric fabrics in the Shikoku Shimanto Belt is still preliminary, but results suggest northward underthrusting. These preliminary results are consistent with similar studies along the strike and with independent data that suggest left-lateral oblique convergence in the Late Cretaceous.

5-1-1:   Pillowed basaltic lava of Valanginian age. The basalt originated as part of the ocean floor basement. The hammer shows the scale. Cretaceous Tei Melange, Geisei Village, Kochi Prefecture, Shikoku (13-Figure 7).

5-1-2:   Pillowed basaltic lava (A) and bedded nannofossil-rich limestone (B) of Valanginian age. The limestone is laminated with thin layers of basaltic tuff, indicating that basalts were emplaced above the CCD. Paleomagnetic measurement of both limestone and pillow lava shows equatorial paleolatitude. The hammer shows the scale. Cretaceous Tei Melange, Yasu Town, Kochi Prefecture, Shikoku (13-Figure 7).

5-1-3: Laminated nannofossil-rich limestone (A) interbedded with basaltic tuff (dark layers, B). Valanginian age. A marker pen shows the scale. Awa Melange, Susaki City, Kochi Prefecture, Shikoku (11-Figure 7).

5-1-4: Bedded ribbon radiolarian cherts of mostly Lower Cretaceous age. Outcrop width is about 7 m. Cretaceous Yokonami Melange, Tosa City, Kochi Prefecture, Shikoku (12-Figure 7).

5-1-5: Folded bedded ribbon radiolarian cherts of Albian-Cenomanian age. One-meter-long scale bar. Cretaceous Tei Melange, Geisei Village, Kochi Prefecture, Shikoku (13-Figure 7).

5-1-6: Cenomanian to Turonian pelagic red claystone with chert intercalations. A marker pen at the center shows the scale. Cretaceous Yokonami Melange, Tosa City, Kochi Prefecture, Shikoku (12-Figure 7)

5-1-7:   Coniacian to Santonian varicolored hemipelagic shale. White layers (A) are tuff, and green-red colored layers (B) are hemipelagite. Cretaceous Tei Melange, Yasu Town, Kochi Prefecture, Shikoku (13-Figure 7).

(5-1-8)

1a

2a

3a

1b

2b

3b

(5-1-8)

4a

5a

6a

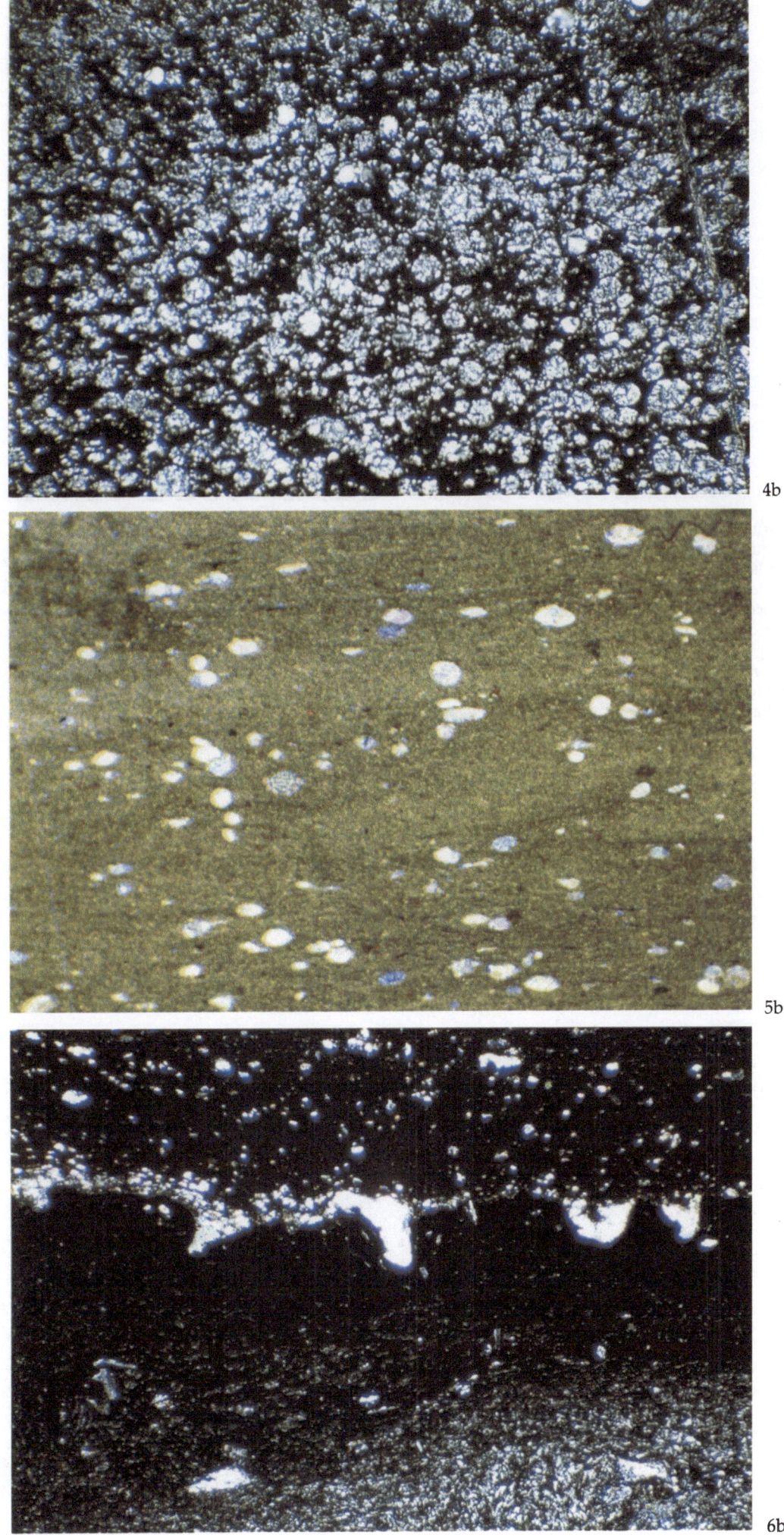

4b

5b

6b

(5-1-8)

7a

5-1-8:   Reconstruction of Cretaceous plate stratigraphy represented by thin-section photomicrographs. Cretaceous Yokonami Melange, Tosa City, Kochi Prefecture, Shikoku (12-Figure 7).
   (1) Sandstone lens in melange matrix. Note cataclastic deformation (A). The sandstone is trench-fill deposits. Campanian age.
   (2) Varicolored hemipelagic shale. Note abundant silt-sized clastic grains mixed with radiolarian tests. Coniacian to Santonian age.
   (3) Pelagic siliceous claystone with scattered radiolarian tests. No coarse clastic particles. Cenomanian-Turonian.
   (4) Bedded radiolarian chert. Note the abundant radiolarian tests and lack of coarse clastic particles. Aptian-Albian age.

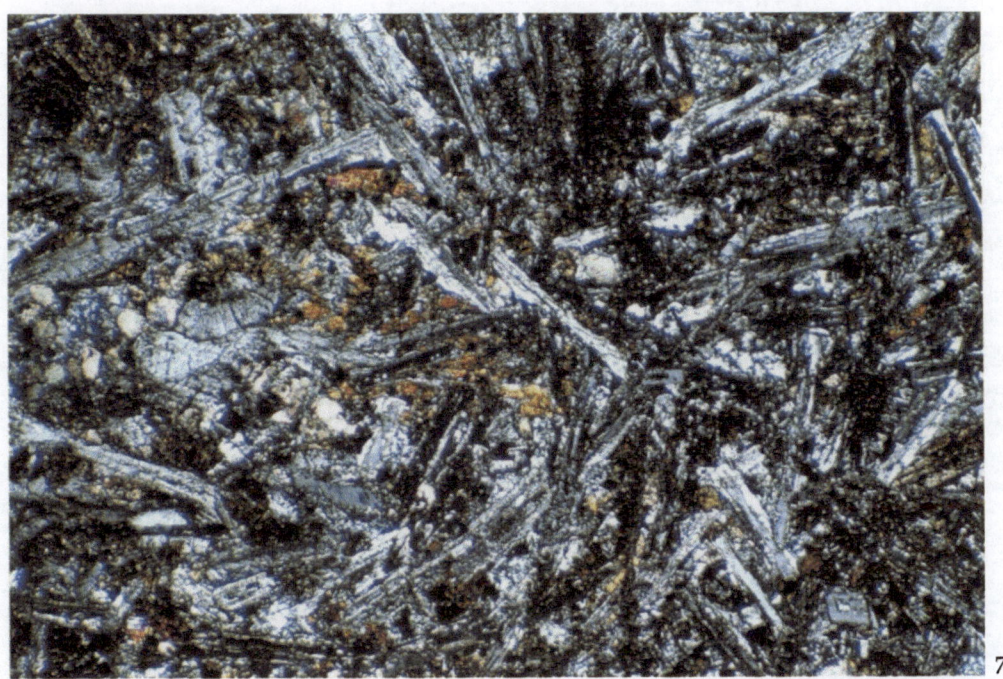

7b

(5) Nannofossil-rich limestone with scattered radiolarian tests. Valanginian age.
(6) Quenched outermost surface of pillow lava (A) and overlying ferro-manganiferous deposits.
(7) Basalt with "swallow tail" lath-shaped plagioclase. Cross nicol on the right side and open nicol
      on the left side. Scale is 2.5 mm.

5-2-1: Coastal outcrop of the Cretaceous Tei Melange, Sumiyoshi Beach, showing the lithological imbrication
         of oceanic plate stratigraphy. (A) is Campanian trench-fill sandstone beds. (B) is a Campanian
         argillaceous melange matrix which was deposited in outer marginal trench environments. (C) is a
         sandstone lens in the argillaceous matrix. (D) is Coniacian to Santonian varicolored shale. (E) is
         Cenomanian to Turonian red pelagic claystone. (F) is Lower Cretaceous radiolarian ribbon cherts. (G)
         is Valanginian pillowed basaltic lava. (H) is a fault. Human figures show the scale. Yasu Town, Kochi
         Prefecture, Shikoku (13-Figure 7).

(5-2-4)

(5-2-2)

5-2-2:
Steeply dipping radiolarian chert beds (50 m across), which are part of a 250-meter chert sequence. Detailed biostratigraphic studies of this sequence (see Figure 9) have shown that the sequence was originally only about 70-100 m thick. Thickening occurred as the chert sequence was imbricated during accretion and underplating. Yokonami Melange, Susaki City, Kochi Prefecture (12-Figure 7).

5-2-3:
Contact between basaltic rock (A) and Lower Cretaceous chert (B). Note black seam of shale between two lithologies (C). This shale is of Coniacian to Campanian age, being apparently younger than rocks above the contact and is a part of the matrix of the melange. A human figure shows the scale. Cretaceous Awa Melange, Kochi Prefecture, Susaki City, Shikoku (11-Figure 7).

5-2-4:
Coniacian-Santonian varicolored shale (A) imbricated with Valanginian basaltic layers above (B) and below (C). A hammer shows the scale. Cretaceous Tei Melange, Geisei Village, Kochi Prefecture, Shikoku (13-Figure 7).

(5-2-3)

5-2-5: Tectonic contact between Cenomanian radiolarian chert-pelagic shale (A) and Campanian black shale-rich melange matrix (B). Note intermixing of black shale and greenish-reddish shale which was originally a part of chert interbedding (C). A hammer shows the scale. Cretaceous Tei Melange, Geisei Village, Kochi Prefecture, Shikoku (13-Figure 7).

5-2-7: Lower Cretaceous chert lenses (A) and basaltic tuff (C) embedded in Coniacian-Santonian varicolored shale (B). The hammer shows the scale. Cretaceous Tei Melange, Yasu Town, Kochi Prefecture, Shikoku (13-Figure 7).

5-2-6:
Tectonic contact between basaltic rocks (A) and melange matrix (B). Note the absence of concentrated shear between the two rock types, and the abundant sandstone lenses (C) within the matrix. A hammer shows the scale. Cretaceous Okitsu Melange, Kubokawa Town, Kochi Prefecture, Shikoku (9-Figure 7).

5-2-8: Albian-Cenomanian chert lens (A) in Campanian melange matrix. The hammer shows the scale. Cretaceous Tei Melange, Geisei Village, Kochi Prefecture, Shikoku (13-Figure 7).

5-2-9: Sandstone boudines (A) with necking surrounded by Campanian melange matrix. Scale bar is 1 m long. Cretaceous Kure Melange, Nakatosa Town, Kochi Prefecture, Shikoku (10-Figure 7).

5-3-2: Example of S-C fabrics and R1 shear on the surface of a slabbed hand specimen. Aligned lenses are mostly hemipelagic varicolored shale. Direction of shear is shown by the arrow. Scale is 1 cm. Cretaceous Okitsu Melange, Kubokawa Town, Kochi Prefecture (9-Figure 7).

5-3-1:
Discontinuous shear planes "C" and folia-tion planes "S" (or S-C fabrics) that indi-cate direction of shear in this view. This would yield a north-directed subduction vector after reconstruction. A hammer shows the scale. Cretaceous Kure Me-lange, Nakatosa Town, Kochi Prefecture, Shikoku (10-Figure 7).

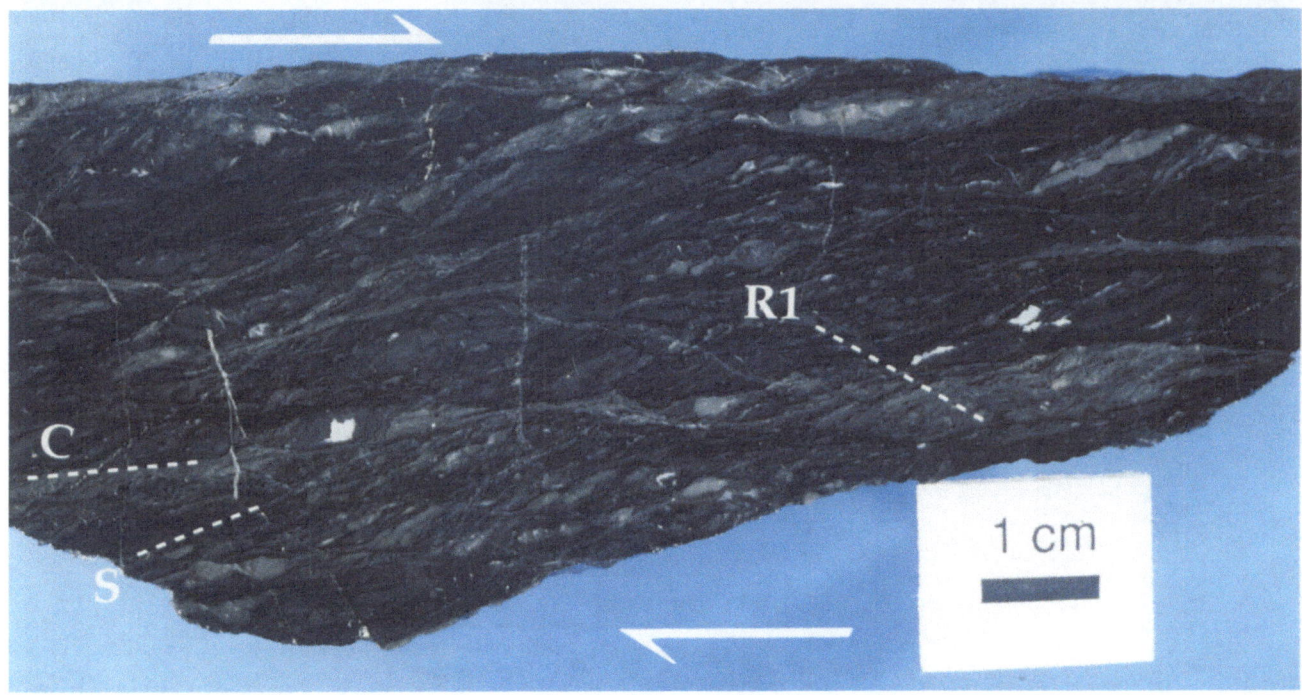

5-3-3: S-C fabrics and R1 shear on the surface of a slabbed hand specimen. Light-colored lenses are mostly hemipelagic varicolored shale and chert. Direction of shear is shown by the arrow. Scale is 1 cm. Cretaceous Okitsu Melange, Kubokawa Town, Kochi Prefecture, Shikoku (9-Figure 7).

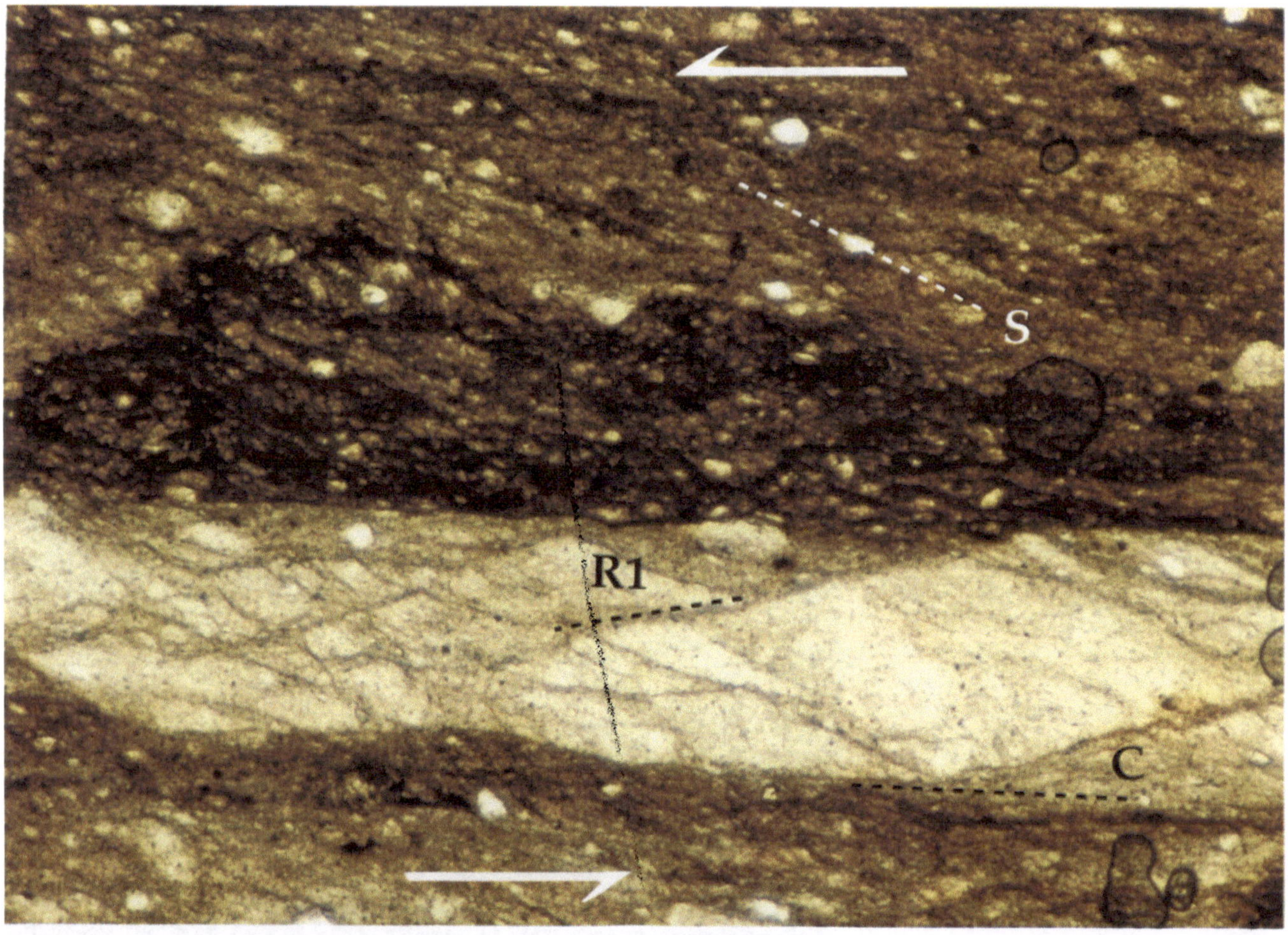

5-3-4:   Photomicrograph showing S-C fabric and R1 shear within varicolored shale lenses.   Open nicol.
Diameter of the photo is 2.5 mm.   Direction of shear is shown by the arrow.   Cretaceous Okitsu
Melange, Kubokawa Town, Kochi Prefecture, Shikoku (9-Figure 7).

5-3-5:   Photograph parallel to foliation surface showing lineations (A) due to alignment of inclusions.   R1
shear steps (B) normal to the lineations are also present.   Cretaceous Okitsu Melange, Kubokawa Town,
Kochi Prefecture, Shikoku (9-Figure 7).

5-3-6: Lenticular Lower Cretaceous chert block (A) with cataclastic tails (B) surrounded by Campanian argillaceous shale matrix. Scale bar is 15 cm. Cretaceous Awa Melange, Susaki City, Kochi Prefecture, Shikoku (11-Figure 7).

5-3-7: Lenticular inclusion of chert block (A) in shale matrix. Note calcite- and quartz-filled shear surfaces in the matrix and extensional fractures in the blocks. Light-colored lenses (B) are composed of hemipelagic varicolored shale. Direction of shear is shown by the arrow. One-centimeter scale. Cretaceous Okitsu Melange, Kubokawa Town, Kochi Prefecture, Shikoku (9-Figure 7).

74

5-3-8: Lenticular block of basalt with abundant quartz-filled veins. 15-cm scale bar. Cretaceous Okitsu Melange, Kubokawa Town, Kochi Prefecture, Shikoku (9-Figure 7).

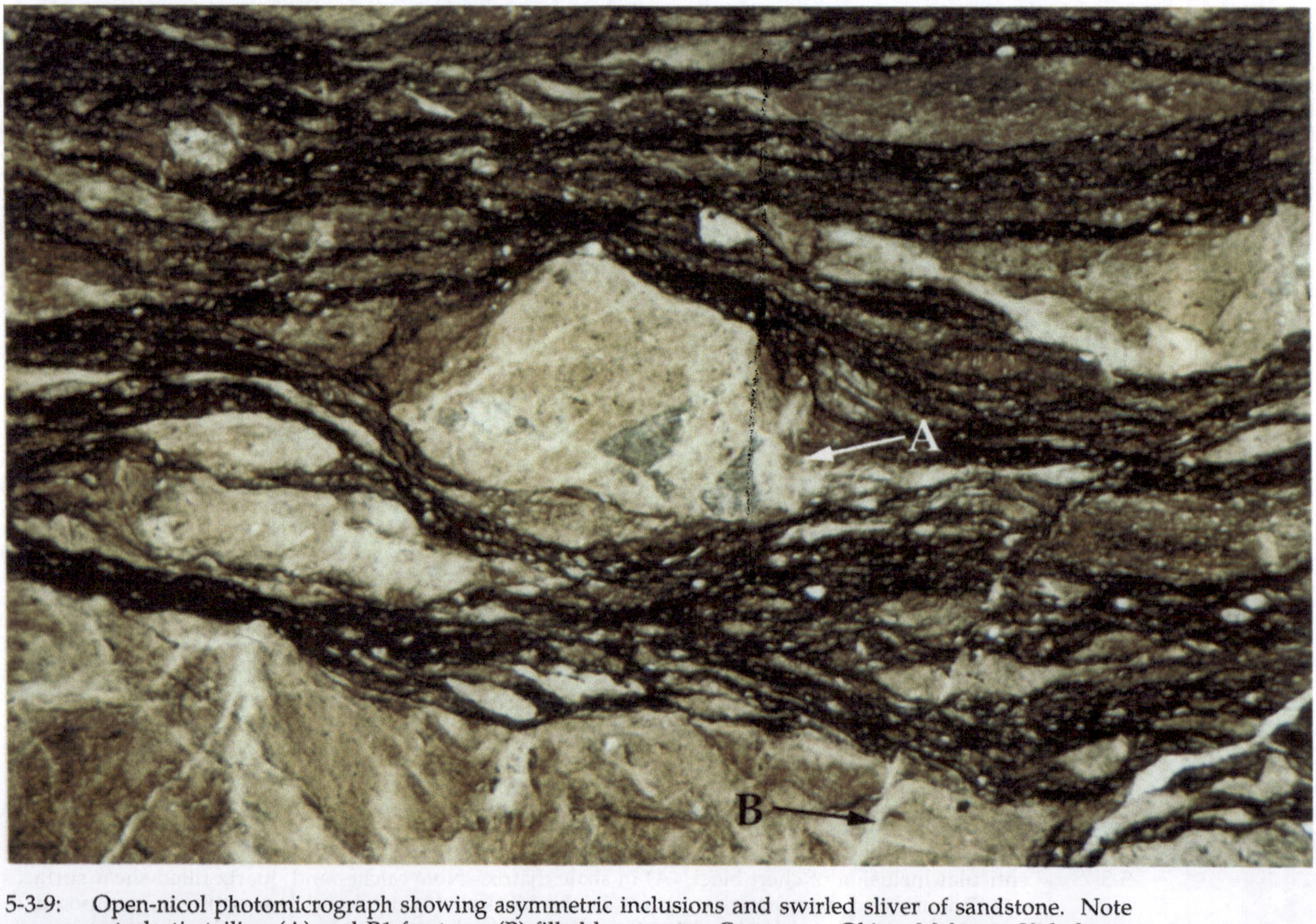

5-3-9: Open-nicol photomicrograph showing asymmetric inclusions and swirled sliver of sandstone. Note cataclastic tailing (A) and R1 fractures (B) filled by quartz. Cretaceous Okitsu Melange, Kubokawa Town, Kochi Prefecture, Shikoku (9-Figure 7).

5-3-10: Rhomboidal sandstone blocks showing various degrees of fracturing and abrasion. 15-cm scale bar. Cretaceous Kure Melange, Nakatosa Town, Kochi Prefecture, Shikoku (10-Figure 7).

5-3-11: Tectonically abraded sandstone block. The hammer shows the scale. Cretaceous Tei Melange, Yasu Town, Kochi Prefecture, Shikoku (13-Figure 7).

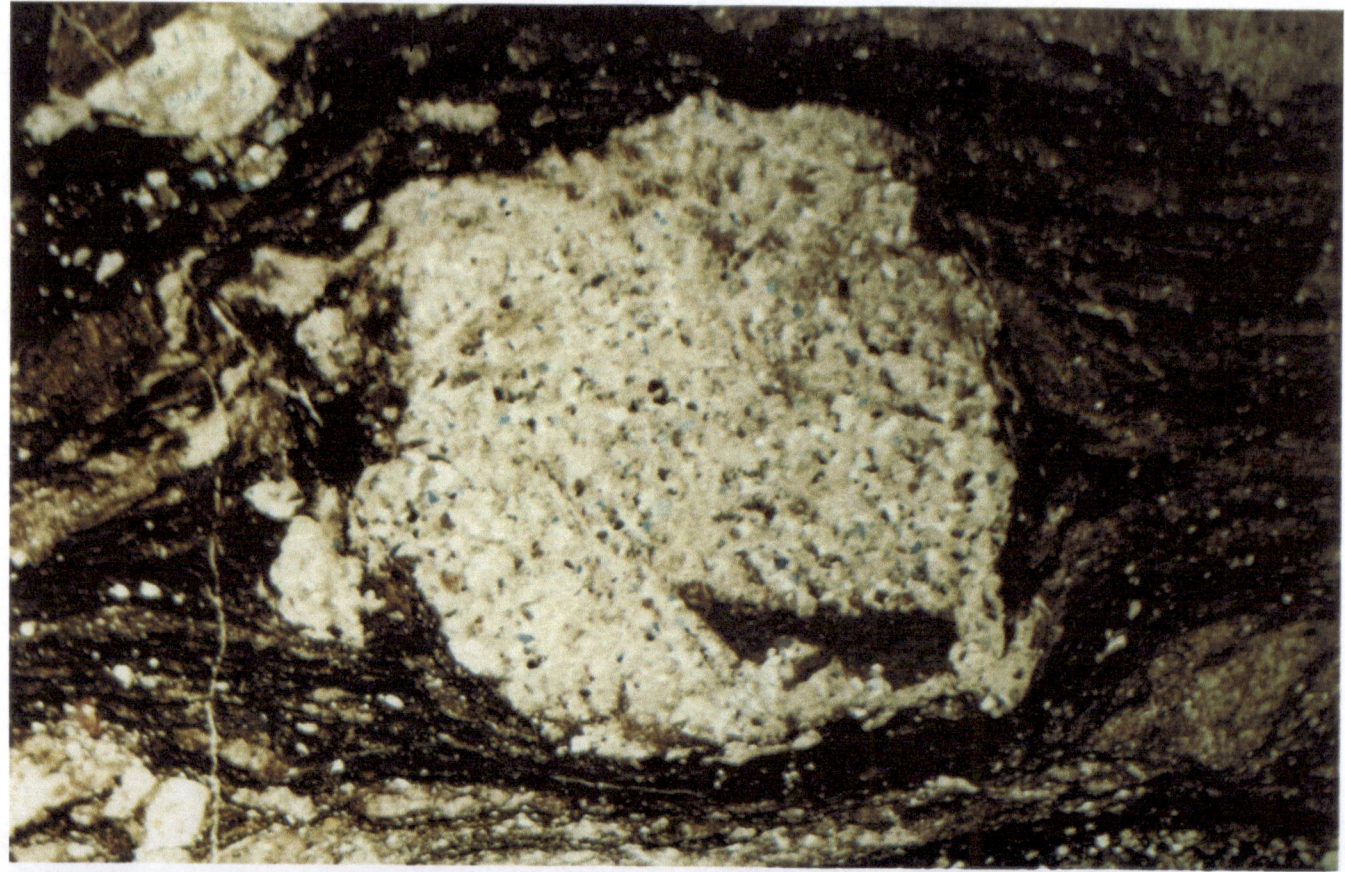

5-3-12: Tectonically abraded sandstone block with asymmetric foliation. Cretaceous Kure Melange, Nakatosa Town, Kochi Prefecture, Shikoku (10-Figure 7).

5-3-13: Photomicrograph of abraded sandstone inclusion in a shale matrix. The inclusion has rotated and involved the surrounding shale matrix. Diameter of the photo is 2.5 mm. Cretaceous Okitsu Melange, Kubokawa Town, Kochi Prefecture, Shikoku (9-Figure 7).

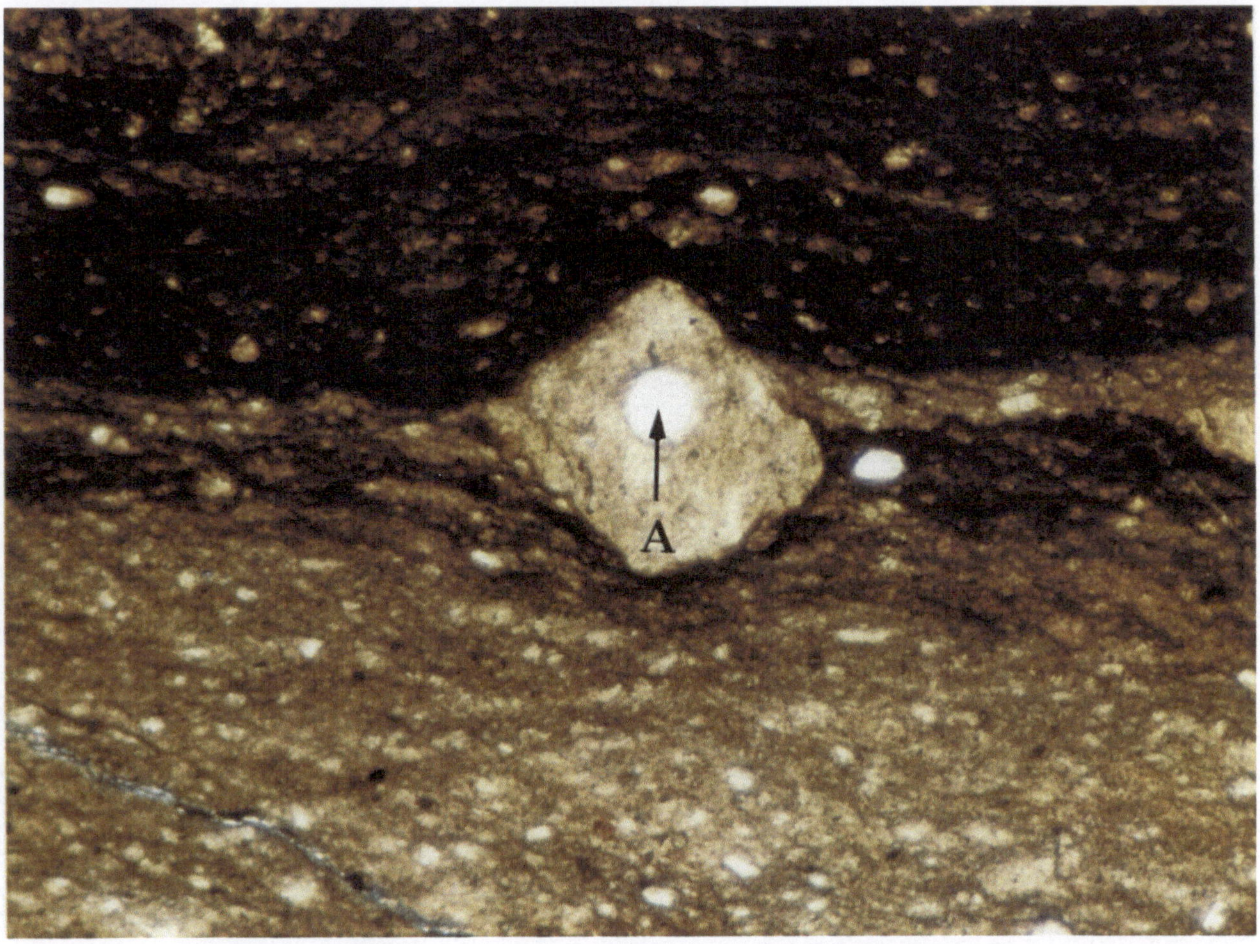

5-3-14: Small chert block with asymmetric cataclastic tail. Note that a radiolarian test (A) is well preserved. Diameter of the photo is 2.5 mm. Cretaceous Okitsu Melange, Kubokawa Town, Kochi Prefecture, Shikoku (9-Figure 7).

5-3-15: Highly mixed asymmetric chert and basalt blocks in a shale matrix. Blocks are asymmetrically swirled and disrupted, indicating a direction of shear. The angle between C and S surfaces is small. Later shear bands (A) cut the foliation. Scale bar is 15 cm. Cretaceous Okitsu Melange, Kubokawa Town, Kochi Prefecture, Shikoku (9-Figure 7).

5-3-16: Highly mixed chert and varicolored shale blocks and lenses in a shale matrix. The angle between C and S surfaces is small. One-centimeter scale bar. Cretaceous Okitsu Melange, Kubokawa Town, Kochi Prefecture, Shikoku (9-Figure 7).

5-3-17: Highly disrupted broken formation in melange zone. Note disrupted small scale folds and sandstone lenses. A pencil shows the scale. Cretaceous Belt, Akaishi Mt., Shizuoka Prefecture, Honshu (10-Figure 4).

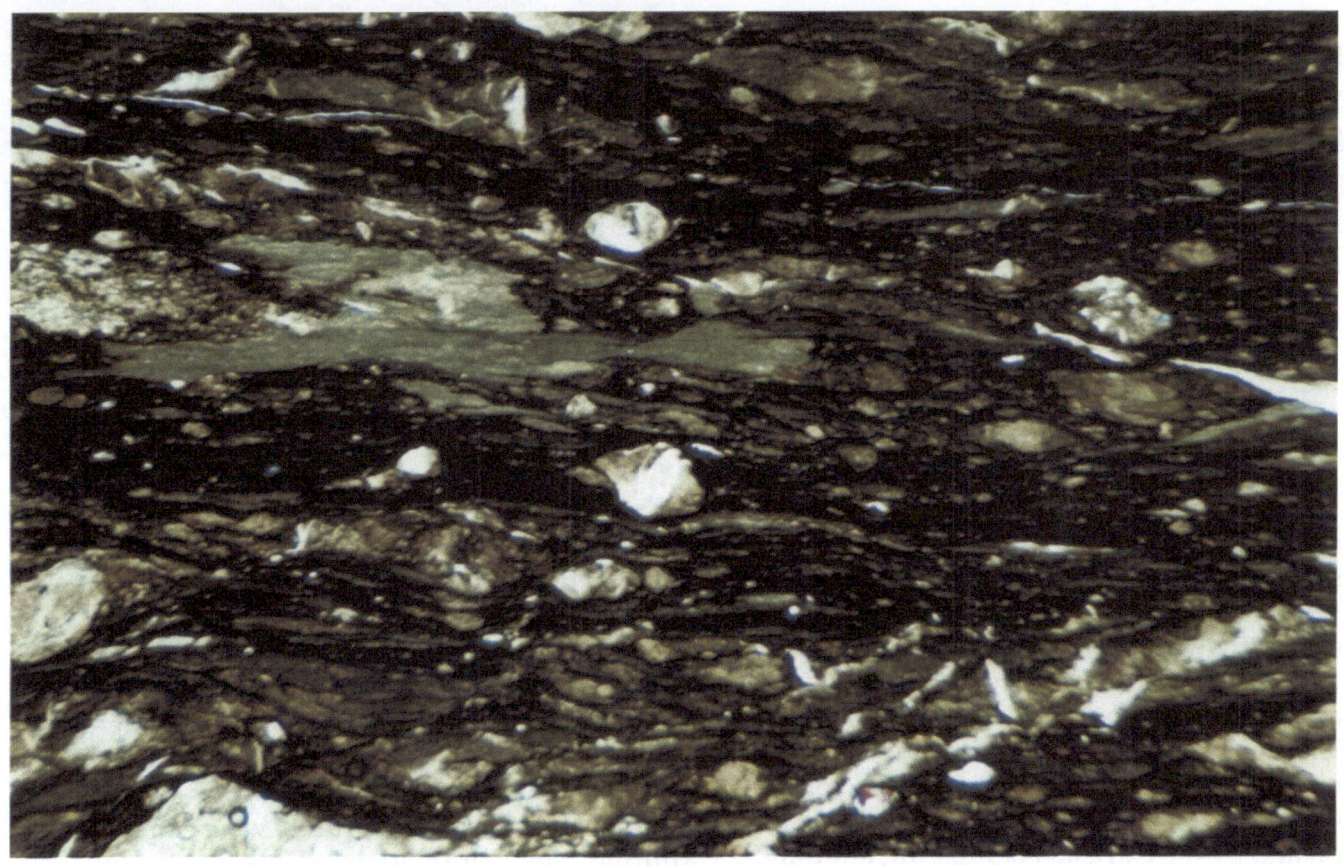

5-3-18: Completely disrupted broken formation in the melange zone. Note abraded inclusions and well-developed foliation oriented approximately parallel to the pencil. Cretaceous Belt, Akaishi Mt., Shizuoka Prefecture, Honshu (10-Figure 4).

5-3-19: Photomicrograph showing highly sheared melange fabrics. The angle between C and S surfaces is close to zero. Note slivers of disrupted varicolored shale beds that are parallel to the foliation. Diameter of the photo is 2.5 mm. Cretaceous Okitsu Melange, Kubokawa Town, Kochi Prefecture, Shikoku (9-Figure 7).

6 COVER SEQUENCES AND NEAR-TRENCH IGNEOUS ROCKS

6-1 Cover Sequences

Sedimentary sequences that overlie, or cover, the accreted turbidite sediments occur as two probably gradational suites: hemipelagic sediments and clastic basin fill. The hemipelagic cover is composed of varicolored to black shale that is distinctively younger than any of the nearby turbiditic sediments. These sequences are also typically less deformed than the accreted sequences. In contrast, the clastic sedimentary sequences are much coarser-grained and often more regionally extensive than the hemipelagic sequences. The clastic cover sequences also typically occur as basin-fill sequences. One of the best examples of a clastic cover sequences is the Misaki Group in Kochi Prefecture (Katto and Taira, 1979) and the Lower Miocene sequence in the Kii Peninsula, Honshu. In these deposits, the clastic sequence starts with a basal olistostrome deposit that contains angular blocks of lithified sandstone and shale (Photos 6-1-1 to 6-1-3). This unit grades abruptly upward into a shale-rich basinal facies which in turn grades upward into an offshore and nearshore sandy facies (Photos 6-1-4 to 6-1-7). This facies is then overlain by tidal to fluvial sandy facies (Photo 6-1-8). Mud diapirism has also been documented in some of the slope basin deposits (Photo 6-1-9).

6-2 Near-Trench Igneous Event: Subduction of Young Ocean Crust or Subduction Initiation

An episodic thermal event affected many parts of the Shimanto Belt and southwest Japan in the Middle Miocene. The thermal event is recorded by short-lived and widespread igneous activity throughout southwest Japan and by the emplacement of a suite of anomalous near-trench igneous rocks. The near-trench rocks include basalts, basaltic andesites and granitoids (Photos 6-2-1 to 6-2-4). Elsewhere in Japan the event was marked by caldera formation and by the extrusion of high-Mg andesitic lavas 14 to 12 Ma in age. Although the origin of this anomalous thermal event remains unresolved, it may have been related to subduction

of the newly formed and relatively "hot" Shikoku Basin lithosphere, to the initiation of subduction along southwest Japan in the Middle Miocene, or to both unusual events.

The Cape Muroto and Cape Shionomisaki areas expose several beautiful examples of near-trench igneous rocks including mafic to acidic lithologies (Miyake, 1988). Detailed field studies have also shown that the gabbroic intrusive suites in Cape Muroto were emplaced during the deformation of host sediments, indicating that near-trench igneous activity occurred contemporaneously with accretion and subduction of the spreading center. The

6-1-1: Wave-cut terrace surface of olistostrome bed. Note dispersed sandstone blocks, most of which are in place. Upper Oligocene-Lower Miocene Muro Group, Sarashikubi, Susami Town, Wakayama Prefecture, Kii Peninsula (9-Figure 4).

Muroto Peninsula area also exposes a sandy slope basin (or cover) sequence (the Shijujiyama Formation) that contains abundant pillowed lava and volcanic breccias of basaltic andesite. This basin therefore represents an unusual example of a volcanic-rich slope basin sequence that formed as young lithosphere was subducted about 15 Ma.

6-3 Metamorphic Xenoliths in the Near-Trench Granitic Rocks

Middle Miocene granitic intrusions in the Shimanto Belt also contain an unusual suite of metamorphic xenoliths (Photo 6-3-1). Several retrieved xenoliths preserve a high grade of metamorphism (e.g., granulite facies) that is interpreted to have formed before the xenoliths were incorporated into the intrusive rock. These inclusions therefore indicate that the deeper levels of the Shimanto Belt have been metamorphosed to substantial temperatures (Komatsu et al., 1991).

82

6-1-2:
Graded debris flow deposit. Note angular lithified blocks of sandstone dispersed in shaley matrix. The hammer shows the scale. Upper Oligocene-Lower Miocene Muro Group, Sarashikubi, Susami Town, Wakayama Prefecture, Kii Peninsula (9-Figure 4).

6-1-3:   Basal olistostrome bed of Lower Miocene Misaki Group. The bed contains angular sandstone blocks, often joint-surfaced, dispersed in shale matrix and is interpreted to be a large-scale debris flow deposit. The hammer shows the scale. Kakumi, Tosashimizu City, Kochi Prefecture, Shikoku (5-Figure 7).

6-1-4:   Well-bedded offshore storm deposits of the lower part of the Misaki Group. The Misaki Group shows an overall coarsening and shallowing upward sequence from offshore to nearshore deposits. The human figure shows the scale. Tosashimizu City, Kochi Prefecture, Shikoku (4-Figure 7).

6-1-5:   Well-bedded sandstone and shale sequence with hammocky cross-bedded sandstone layers indicating sandbar deposits. The hammer shows the scale. Misaki Group, Tosashimizu City, Kochi Prefecture, Shikoku (4-Figure 7)

(6-1-7)

(6-1-6)

6-1-6: Hammocky cross-laminated sandstone showing shallow water sandbar depositional environments. Human figures show the scale. Misaki Group, Tatsukushi, Tosashimizu City, Kochi Prefecture, Shikoku (3-Figure 7).

6-1-7: Convolute lamination within nearshore sandbar deposits. Note that the upper part of the convolution is truncated by an overlying sand bed, indicating that convolution occurred before the deposition of the overlying beds. Extensive convolution within sandbar deposits may indicate occasional strong seismic activity in growing accretionary prism. The camera lens cap shows the scale. Misaki Group, Tatsukushi, Tosashimizu City, Kochi Prefecture, Shikoku (3-Figure 7).

6-1-8: Bimodal directions of coarse-grained sandstone cross-beds indicating tidal depositional environments. The hammer shows the scale. The uppermost part of the Misaki Group, Tatsukushi, Tosashimizu City, Kochi Prefecture, Shikoku (3-Figure 7).

(6-1-8)

(6-1-9)

(6-2-1)

(6-2-2)

6-1-9: Surface of mud diapir intruding into offshore sandbar deposits. The hammer shows the scale. Miocene Tanabe Group, Hikigawa Town, Wakayama Prefecture, Kii Peninsula (7-Figure 4).

6-2-1: Helicopter aerial photo of the coastal terrace exposures of Cape Muroto. Middle Miocene (15 Ma) Gabbroic intrusive (A) is in contact with Upper Oligocene-Lower Miocene sandstone and shale sequence (B). This gabbro is about 150 m thick at maximum. The gabbroic intrusives found in this area have been interpreted as igneous activity due to subduction of the active spreading center of the Shikoku Basin. Muroto City, Kochi Prefecture, Shikoku (18-Figure 7).

6-2-2: Outcrop of gabbro intrusive at Cape Muroto. Paleomagnetic data and texture of the pegmatite-like pocket indicate that the igneous body was an originally subhorizontally intruded sill which was later tilted to near vertical dip. The contact zone with the sediment also shows the evidence for intrusion into semi-lithified sediments such as irregular contact surface and hydrothermal alteration. Muroto City, Kcohi Prefecture, Shikoku (18-Figure 7).

6-2-3:   Pillow lavas of basaltic andesite which compose the basal part of a Miocene slope basin. Shijujiyama Formation, Ohbae, Muroto City, Kochi Prefecture, Shikoku (19-Figure 7).

6-2-4:   Middle Miocene granitic rock exposed at Cape Ashizuri. A widespread granitic rock intrusion in the Shimanto Belt indicates an unusual near-trench heating event. Tosashimizu City, Kochi Prefecture, Shikoku (6-Figure 7).

6-3-1:   A xenolith of banded gneiss of psammatic rock origin. The xenolith is metamorphosed to granulite facies (about 7 kb, 800°C). Bands are composed of a quartz-feldspathic layer and a hypersthene/ garnet-bearing biotite-rich layer and are folded with axial plan schistosity. Host granitic rock is 14 Ma tourmaline-biotite granodiorite. Takatsukiyama, Ehime Prefecture, Shikoku. Photo courtesy of M. Komatsu (1-Figure 7).

7 CORRELATIONS WITH THE SANBAGAWA BELT AND EVIDENCE FOR PROGRESSIVE GROWTH OF THE PRISM

7-1 Shimanto Belt and Sanbagawa Belt

Recent structural and radiometric studies have suggested that the Shimanto Belt grades landward into part of the Sanbagawa Belt. The Sanbagawa Belt is a regional-scale belt of Cretaceous high-pressure and low-temperature metamorphic rocks that are present for hundreds of kilometers along southwest Japan. Locally, the Sanbagawa Belt preserves paleo-pressures equivalent to burial deeper than 30 km. The recently proposed correlation between part of this belt and the Shimanto Belt is therefore significant because it suggests that the Late Cretaceous units may preserve a relatively complete accretionary system, from very shallow levels (e.g., in the Early Tertiary to Cretaceous Shimanto Belt) to very deep levels (e.g., the high-pressure Sanbagawa Belt).

The proposed equivalent part of the Sanbagawa Belt is the Oboke Sandstone, which represents the lowest structural unit within the Sanbagawa Belt and consists of a sequence of sand and shale with local conglomeratic horizons that has undergone low-grade metamorphism (greenschist facies) (Isozaki and Itaya, 1991). The clasts of conglomerate include quartzite, shale and granitoid (Photos 7-1-1 and 7-1-2). The Oboke Sandstone therefore is lithologically very similar to the Shimanto Belt flysch sequences. The clasts have been flattened parallel to the schistosity, and a stretching lineation is locally well developed. Quartz-filled pressure shadows around the clasts are locally asymmetric.

Evidence for the progressive growth and uplift of the Cretaceous accretionary prism is also provided by the increased volume of basaltic rocks with an increased grade of metamorphism (Isozaki et al., 1990). These results suggest that oceanic materials (e.g., basaltic rocks) are more likely to be underplated in the deeper levels of the prism. Ogawa and Taniguchi (1989) have also shown that the composition of the accreted basaltic rocks changes with metamorphic grade: seamount affinities are more abundant in lower grade rocks and ocean floor affinities are more abundant in higher grade rocks. Again, these results suggest that deeper stratigraphic levels on the subducting plate are accreted (or underplated) at deeper structural levels.

These lines of evidence indicate that underplating of materials is the dominant mechanism for the growth of accretionary prisms including the exhumation of high-pressure-type metamorphic rocks. The geologic evidence for erosion of the Sanbagawa Belt has been documented in the schist clasts of the Eocene Kuma Group in Shikoku, which lies directly on the Sanbagawa Belt (Photo 7-1-3). This suggests that about 20 Ma was required for the Sanbagawa Belt to be exhumed.

7-2 Uplift of the Accretionary Prism

The final stage of deformation in the Shimanto Belt includes the development of brittle structures at all levels. Brittle structures can be recognized at nearly every outcrop in the Shimanto Belt and appear to be responsible for much of the present-day topography. The Muroto Flexure and the east-trending Shiina-Narashi fault on the Muroto Peninsula provide two of the best-studied examples. The Shiina-Narashi fault represents the contact between the Eocene to Oligocene and Miocene Shimanto Belts on the north, whereas the Muroto Flexure is a regional-scale, north-trending cross-fold. The flexure appears to have rotated older, accretion-related, regional-scale fold axes from subhorizontal plunges on the west coast to subvertical plunges on the east coast of the Muroto Peninsula. It also appears to be a relatively recent structure as it affects both the Eocene to Oligocene rocks and the Oligocene to Miocene rocks to the south (Figure 6). Sugiyama (1989) has also suggested that the flexure may still be active and related to the present-day oblique convergence between the Philippine Sea and Eurasia plates in southwest Japan.

Sugiyama (1989) and Okamura (1990) have also recently compiled the distribution of major, recently active faults along southwest Japan and have proposed that the two conspicuous promontories of Shikoku Island reflect motion along oblique, high-angle thrust faults that cross-cut the structural grain of the Shimanto Belt (e.g., the Muroto Flexure discussed above).

This late-stage uplift pattern is also consistent with the paleotemperature patterns revealed by vitrinite reflectance, illite crystallinity (DiTullio et al., in press) and fission track analyses. These data show that most of the Cretaceous Shimanto Belt (Northern Belt) experienced temperatures within the zircon annealing zone (190–260°C) and that parts of the melange zones experienced temperatures above 260°C. In contrast, most of the Tertiary Shimanto Belt experienced maximum temperatures within the apatite annealing zone (60–120°C), except for some of the Late-Oligocene-Miocene rocks which were heated by an anomalous near-trench igneous event (discussed below). Finally, the fission track age data suggest that both the Cretaceous and Tertiary Shimanto Belts were uplifted coevally and within the last 10 Ma (Hasebe et al., in press).

Direct evidence for rapid uplift of the Shimanto accretionary prism can be suggested by analysis of coastal marine terrace. The altitude of the middle terrace (M surface, 120,000 yrs old) is about 190 m and that of the lower terrace (L surface, 6,000 yrs old) is about 10 m at Cape Muroto where the coastal uplift records the highest rate (Photos 7-2-1 to 7-2-3). These data provide a mean uplift rate of 2 m/1000 yrs. This suggests that the magnitude of uplift in 10 Ma is about 20 km. Although the exact depth of burial in the Cretaceous and Tertiary Shimanto Belts is not known, we estimate it to be 5 to 10 km or so, and that it took about 10 Ma to be exhumed. The average depth of the Sanbagawa Belt metamorphism has been estimated as 20 to 30 km. This depth divided by 20 Ma exhumation duration provides 10 to 15 km per 10 Ma. It is remarkable that these crude estimations of exhumation rate all give a narrow range of values (5 to 20 km per 10 Ma).

7-1-1: The photo shows a conglomerate outcrop of the Ohboke Sandstone exposed along the river beds of the Ohboke Gorge of the Shikoku Mountain range. The clasts have been flattened parallel to the schistosity, and there is a stretching lineation that is locally well developed. The mean aspect ratio of the clasts is about 3/10. The clasts are three types of quartzite: shale, acidic volcanic rock and granitoid. A recent theory is that the Ohboke Sandstone may represent a time-equivalent sequence of the Shimanto Belt which has been subcreted underneath the older accretionary complex. The hammer shows the scale. Ohboke Gorge, Tokushima Prefecture, Shikoku (5-Figure 4).

7-1-2: View of the surface of the Ohboke Sandstone parallel to the schistosity. Note flattened acidic rock clast. Ohboke Gorge, Tokushima Prefecture, Shikoku (5-Figure 4).

7-1-3: Conglomerate outcrop of the Eocene Kuma Group which is one of the oldest sedimentary records of the denudation of the Sanbagawa Belt. Note the large size of schist clasts in alluvial fan deposits. The hammer shows the scale. Iwayaji, Ehime Prefecture, Shikoku (3-Figure 4).

7-2-1: Oblique helicopter aerial photo of Cape Muroto. Note the well-developed marine terrace. The prominent flat topographic surface is 150,000 yrs old (Shimosueyoshi Surface) and was produced by the last episode of maximum transgression. Muroto City, Kochi Prefecture, Shikoku (18-Figure 7).

7-2-2: Oblique helicopter aerial photo of the well-developed marine terrace surface at Cape Gyodo. Note the development of two main terrace surfaces. The upper surface (Middle Terrace) is 150,000 yrs old and the lower surface (Lower Terrace) is about 1,000 yrs old. Cape Gyodo, Muroto City, Kochi Prefecture, Shikoku (17-Figure 7).

7-2-3: The uplifted wave cut notch, which is now 4.1 m above sea level. Note the fossil calcareous secretion being pointed to by the person. The C^{14} date of this calcareous remains is 830 yrs. Cape Muroto, Muroto City, Kochi Prefecture, Shikoku (18-Figure 7).

(7-2-1)

(7-2-2)

(7-2-3)

8 SEDIMENT DEFORMATION IN THE SHALLOW LEVELS OF AN ACCRETIONARY PRISM: EVIDENCE FROM MIURA AND BOSO PENINSULAS

The Izu Collision Zone (ICZ) encompasses the region of the Neogene to Recent arc-arc collision between the main island of Japan, or Honshu Island arc, and the Izu-Ogasawara (Bonin) Island arc immediately southwest of the Boso Peninsula (Figure 2). This area represents the junction between three active plate convergence zones: the Japan Trench to the northeast, the Izu Trench to the southeast and the Nankai Trough to the southwest. This juncture therefore represents one of the few active trench-trench-trench triple junctions in the world. The present collisional boundary runs west to east through the Suruga Trough and under Mt. Fuji to the Sagami Trough and to the triple junction area to the east (Figure 2).

Reconstruction of these interacting and dynamic plate boundaries around the ICZ suggests that the Izu-Ogasawara arc has been accreted successively through time together with a thick, coarse clastic sequence of trench-fill deposits (Photos 8-1-1 and 8-1-2). The sequence is composed of 3 to 4 km thick, coarsening-upward and shallowing-upward clastic wedge. Accretion probably occurred through a type of crustal delamination or imbrication (Taira et al., 1989; Soh et al., 1991).

The Miura Group (Eto et al., 1987) is an important component of this collisional complex and ranges in age from Late Miocene to Pliocene (10-3 Ma), which encompasses most of the period in which the collision was active. The Miura Group is composed mainly of scoriaceous to pumiceous volcaniclastic sediments derived from the Izu-Ogasawara arc and deposited in a deep-shelf to basinal environment. These sediments are now exposed along the beautiful marine terraces that formed during the late stages of the arc-arc collision. The Miura Group therefore forms part of a Neogene accretionary complex in the ICZ area.

8-1 Deformation and Fluid Flow Recorded in the Miura and Boso Peninsulas

Deformation of the Miura Group is characterized by an unusually abundant number of structures associated with fluidization and dewatering of the sediments. Typical structures include mud-filled veins, clastic intru-

sions and associated breccias and outcrop-scale faults and folds. The Miura Group also preserves a spectacular example of a submarine fluid vent (or hot spring) with an associated fossil assemblage of *Calyptogena* and other deep-sea clams. These unusual deformational and fluid flow structures are discussed separately below.

8-1-1 Mud-Filled Veins (Vein Structure)

Mud-filled veins (previously called "vein structures") are particularly abundant in the Miura Group. Individual veins tend to be relatively planar and are typically only 0.1-0.2 mm wide in the central part of the vein (Photos 8-1-3 and 8-1-4). Several generations of veins have been recognized, however, and the later veins tend to be larger and more widely spaced; many are up to several millimeters in width. Sigmoidal veins have also been recognized locally. The mud-filled veins also often occur in sets or bands in which ten to one hundred subparallel veins are concentrated. The bands are usually only 1 to 10 cm thick and typically form parallel to sedimentary layering. Individual veins are nearly perpendicular to bedding and they are often equidistant from each other, ranging from 1 to 30 mm apart (Ogawa, 1980).

Several observations also suggest that the mud-filled veins formed relatively early in the deformation history of the Miura Group. First, in nearly all cases where cross-cutting relations have been observed the veins are deformed by the outcrop-scale faults. In the few cases where the veins are observed to cut fault zones, the faults also appear to be relatively early. Second, mud-filled veins occur within blocks of sediment caught up in early, chaotic, sediment slide deposits, indicating that the veins formed before these early gravitationally driven deposits. Finally, unpublished X-ray tomographic analyses of mud-filled veins indicate that the veins are denser than the matrix sediment, suggesting porosity reduction during their formation. The mud-filled veins are therefore considered to have formed early in the deformational history of the sediments, and probably at relatively shallow depths.

Mud-filled veins have also been recognized from various forearc trench-slope settings through the Deep Sea and Ocean Drilling Programs (see summary by Lundberg and Moore, 1986). Recently, Kimura et al. (1989) have documented mud-filled veins within a few meters of the seafloor in unconsolidated volcaniclastic muds. Recent drilling in the Mariana and Izu-Bonin forearcs has also revealed the occurrence of vein structures (probably mud-filled veins) within 100 m of the seafloor (Fryer et al., 1990; Taylor et al., 1990).

A number of authors have proposed mechanisms for the origin of mud-filled veins and vein-like structures. Knipe (1986) and Leggett et al. (1987) considered these structures to form in response to gravity-induced downslope failure of partly lithified sediment. Knipe (1986) also noted the possibility that veins may develop as, or be modified into, a sigmoidal geometry as a result of shear parallel to bedding in unconsolidated slope sediments (see also Kimura et al., 1989). Pickering et al. (1990) have suggested that the veins in the Miura Group formed during downslope or gravity-induced extension.

8-1-2 *Slumping, Fluidization and Brecciation*

Evidence of remobilization and fluidization of sediments in the Miura Group is present at a number of outcrops (Photos 8-1-5 to 8-1-8). Many of the remobilized units are composed of scoria or very coarse-grained tuff and, when remobilized, fill cracks and veins ranging in size from only a few millimeters to several meters. As this remobilized material is transported and injected the adjoining wall rock can also be deformed. Locally, this deformation results in a relatively intense interconnected network of cracks, faults and fractures that are then filled with the remobilized scoria or tuff. In some locations this deformation and infilling result in lithologies that are similar to conglomerates or breccias. These lithologies are more accurately described as autobreccia, however, because the brecciation or mixing is interpreted to have formed nearly in situ.

8-1-3 *Faulting*

The well-exposed coastal outcrops in the Miura and Boso Peninsulas also reveal a polyphase history of deformation dominated by the three-dimensional geometry of faults and folds. Many of the faults, however, are also distinctive because they appear to be annealed (Photos 8-1-9 to 8-1-12). That is, the faults are not presently zones of weakness and do not lack cohesion. Instead, the faults appear to have been cemented or lithified to the same degree as the wall rock. These faults therefore are considered to have formed relatively early in the deformation history and prior to complete lithification.

The abundance of outcrop-scale faults with minor displacements has also allowed the application of paleostress techniques. Kakimi et al. (1966) documented progressive stress field changes in the Miura Peninsula. More recently, Angelier and Huchon (1987) have proposed that the major compressional direction changed from NNE-SSW to NNW-SSE in the Miura-Boso Peninsulas. They have also suggested that this change in shortening direction reflects a change in the convergence direction and/or block rotations within the Miura-Boso region during the Izu collision.

8-2 Fossil Fluid Venting and *Calyptogena* Beds

The Miura Peninsula region also exposes several fossilized giant clam colonies of *Calyptogena* (Niitsuma et al., 1989). The fossil colonies occur within the deep-marine (upper bathyal water depths based on benthic foraminifera) volcanic sandstones and conglomerates of the Ikego Formation (Photos 8-2-1 to 8-2-3). Our unpublished field notes show that the *Calyptogena* colonies in this formation are often concentrated in zones where there is also evidence for high fluid pressures and synsedimentary brecciation. For example, the zones often contain numerous clastic dikes and fluidized breccia, and are interbedded with debris flow or breccia deposits. The zones are also often extensively cemented, and some of the cemented zones form shapes suggestive of circular conduits or chimneys. Calcite cements obtained from a shell-bearing volcanic-rich conglomerate also have light carbon isotopic ratios ($\delta^{13}C = -19.7$ to -49.2) (unpublished data by Soh and Taira), suggesting an origin related to biogenic methane and fluid seepage. Finally, some of the fossil shells occur in living positions, indicating that the clams lived near the vent sites. The reconstructed environment is therefore very similar to the mid-slope of the modern Sagami accretionary prism, with clam colonies living near sites of concentrated dewatering, or submarine vents.

8-1-1: Steeply bedded conglomerate sequence of the Pleistocene fan-delta Shiozawa Formation, Ashigara Group. Bedding is shown by (A). The unconformably overlying stratum is a Late Pleistocene tephra sequence. The bulldozer shows the scale. Yamakita Town, Kanagawa Prefecture (11-Figure 4).

8-1-2: Shattered clast in the Pleistocene Shiozawa Formation, Ashigara Group. Faulting is common in the Ashigara Group. Yamakita Town, Kanagawa Prefecture (11-Figure 4).

8-1-3: Mud-filled vein array in the siltstone of the lower Miura Group. Vein-bearing zones are subparallel to the bedding planes, although vein zones often anastomose each other. Veins usually disappear near a coarse-grained region such as a scoriaceous sandstone layer or a burrow filled by sand. Boso Peninsula, Chiba Prefecture (13-Figure 4).

8-1-5:   Slump structure of the Miura Group in the Boso Peninsula. Scale bar is 1 m long. Note the slump folds and isolated sedimentary blocks in the massive tuffaceous siltstone. Thin-bedded tuffaceous layers at the upper part of this photograph are overturned. The slumping appears to be attributed to slope instability due to rapid sedimentation and/or forearc crustal movement. Boso Peninsula, Chiba Prefecture (13-Figure 4).

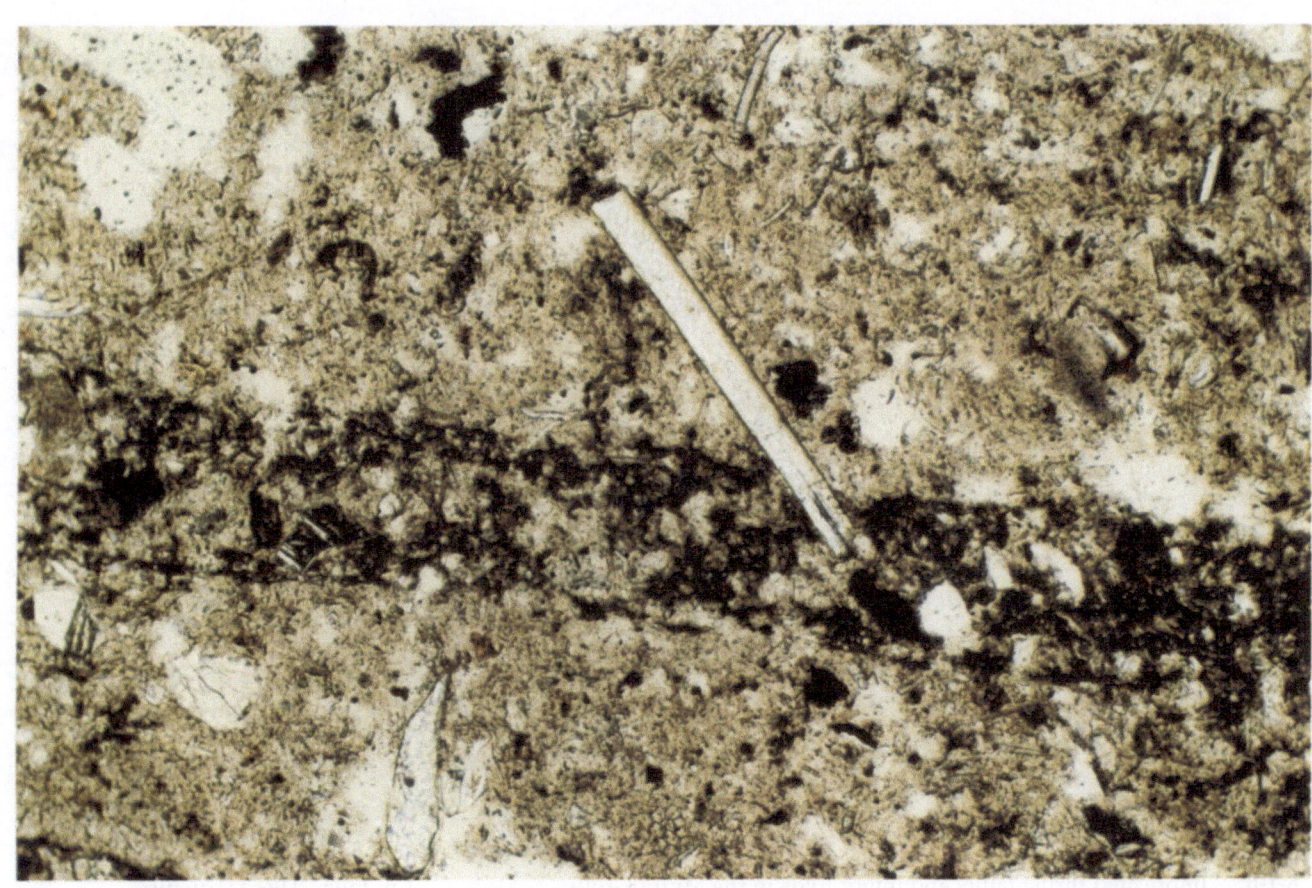

8-1-4: Photomicrograph of mud-filled vein. The vein is composed of finer-grained material than is the surrounding host rock. Sponge spicules appear to be undeformed across the vein wall, suggesting the absence of cataclastic deformation. 2.5 mm is the width of the area in the photo. Boso Peninsula, Chiba Prefecture (13-Figure 4).

8-1-6:   Imbricated thrust faults and minor folds in the Miura Group. Scale bar is 1 m long. These structures are regarded as compressional folding and thrusting at the toe of the slump. The convergence of both folds and faults, which suggests the direction of sliding and downslope movement, is predominantly towards the southeast. Boso Peninsula, Chiba Prefecture (13-Figure 4).

8-1-7:   Chaotic units in the Miura Group. Scale bar (lower left corner) is 1 m long. It comprises several blocks enclosed by healed faults without any matrix. Each block is disrupted and folded sedimentary rock. This structure is interpreted to be an amalgamation of slumped sequences caused by a tectonically unstable environment. Boso Peninsula, Chiba Prefecture (13-Figure 4).

8-1-8:   Clastic sills and dikes in the upper Miura Group distributed in the southern tip of the Boso Peninsula. Matrix of the clastic intrusion is scoriaceous sand derived from the nearby sedimentary layers. Scale bar is 15 cm long. Boso Peninsula, Chiba Prefecture (13-Figure 4).

8-1-9:
Conjugate sets of thrust faults in the Miura Group of the Boso Peninsula. Fault planes are completely annealed. Scale bar is 1 m long. Direction of horizontal maximum compressional stress is estimated to be NW-SE after bedding correction. Boso Peninsula, Chiba Prefecture (13-Figure 4).

8-1-10:
Clastic dikes observed in the siltstone of the Miura Group. Dikes are mainly composed of tuffaceous matrix and siltstone blocks. Note several pieces of dikes separated by healed faults subparallel to the bedding plane. Boso Peninsula, Chiba Prefecture (13-Figure 4).

8-1-11: Thrust zone composed of repetition of the same sedimentary layer at the wave-cut terrace in the Boso Peninsula. Healed fault planes suggest active thrusting before lithification. This zone is observed beyond several hundred meters along NW-SE trending shoreline. Similar structures are also well developed in the Miura Peninsula. Boso Peninsula, Chiba Prefecture (13-Figure 4).

8-1-12: Conjugate sets of normal faults in the Miura Group. Scale bar is 1 m long. Clastic dikes such as that shown in Figure 8-1-10 also occur around this structure. Boso Peninsula, Chiba Prefecture (13-Figure 4).

8-2-1: Shell beds of *Calyptogena*. Reworked shells are found in coarse-grained volcaniclastic sediments of the Pliocene Ikeogo Formation. The Ikeogo Formation is composed mainly of canyon-fill sediments in a trench slope setting. Zushi City, Kanagawa Prefecture (12-Figure 4).

8-2-2: Carbonate cemented chimney-like structure (A) associated with *Calyptogena* shell beds. The structure resembles closely those found on the seafloor. The existence of such cementation indicates the near-vent nature of shell beds. Pliocene Ikeogo Formation, Zushi City, Kanagawa Prefecture (12-Figure 4).

8-2-3: Carbonate geopetal cementation (A) underneath a shell. Early seafloor cementation can be suggested by the layer-parallel geopetal and reworked cemented shell beds in intraformational slump deposits. Pliocene Ikeogo Formation, Zushi City, Kanagawa Prefecture (12-Figure 4).

9 NANKAI TROUGH: A MODERN ANALOGUE

9-1 Tectonic Framework

The Nankai Trough (Figure 2) is the topographic expression of the plate boundary between the Philippine Sea and Eurasian plates. Seismic slip data indicate a convergence south of the island of Shikoku at the rate of 3-4 cm/yr (Seno, 1977), whereas geologic constraints indicate a slightly lower rate of 2-3 cm/yr (Karig and Angevine, 1986). In the area of Shikoku Island the Philippine Sea plate is composed of the Oligo-Miocene Shikoku Basin (Kobayashi and Nakada, 1978; Shih, 1980; Chamot-Rooke et al., 1987) and, because the basin is relatively young, the Nankai Trough is relatively shallow (maximum water depth 4.8 km). The trough is also partially filled by a significant pile of sediments (1 to 2 km thick) (Le Pichon et al., 1987).

In the region of the Nankai Trough there are basically two layers of sediments being subducted beneath the forearc: an upper layer rich in turbidite deposits and a lower stratigraphic succession rich in hemipelagic deposits (Kagami et al., 1986; Taira et al., 1991). The turbidites are being supplied mainly from the Suruga Trough drainage area to the east, especially from the Fuji River in central Japan (Taira and Niitsuma, 1986) at a high rate of deposition. Locally this rate reaches 2000 m/m.y., or more, and is comparable to the present rates of deposition found in the coastal to fluvial environments around Japan. Taira and Niitsuma (1986) have shown that this enormous sediment flux is directly related to the arc-arc collision between the Izu-Bonin island arc and the Honshu (mainland Japan) arc. The basal hemipelagic sediments show a varicolored appearance similar to those observed in the Shimanto Belt.

The region just landward of the Nankai Trough (the Nankai forearc) is composed of the modern accretionary prism, which is partly covered by slope sediments, a trench-slope break and several forearc basins. Deformation of the trench sediments is initiated in a "proto-thrust zone", where progressive, and probably penetrative, thickening of the trench wedge occurs (see seismic reflection profiles in Figure 11; Moore et al., 1990; Karig and Lundberg,

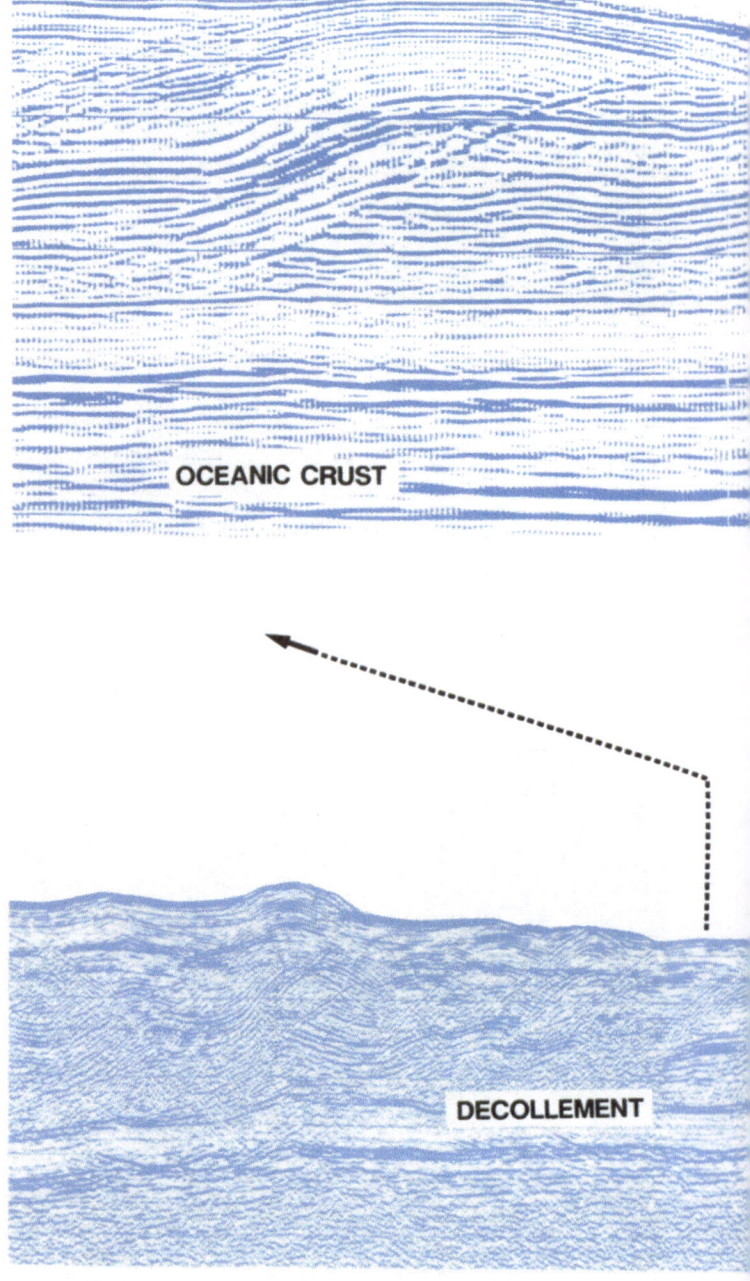

1990).

The frontal thrust zone occurs just landward of the proto-thrust zone and shows a series of regularly spaced (1-2 km spacing) active thrust faults (see Figure 11). ODP Leg 131 drilling results revealed overturned beds associated with the frontal thrust. This strongly supports the concept of an early-stage faulting

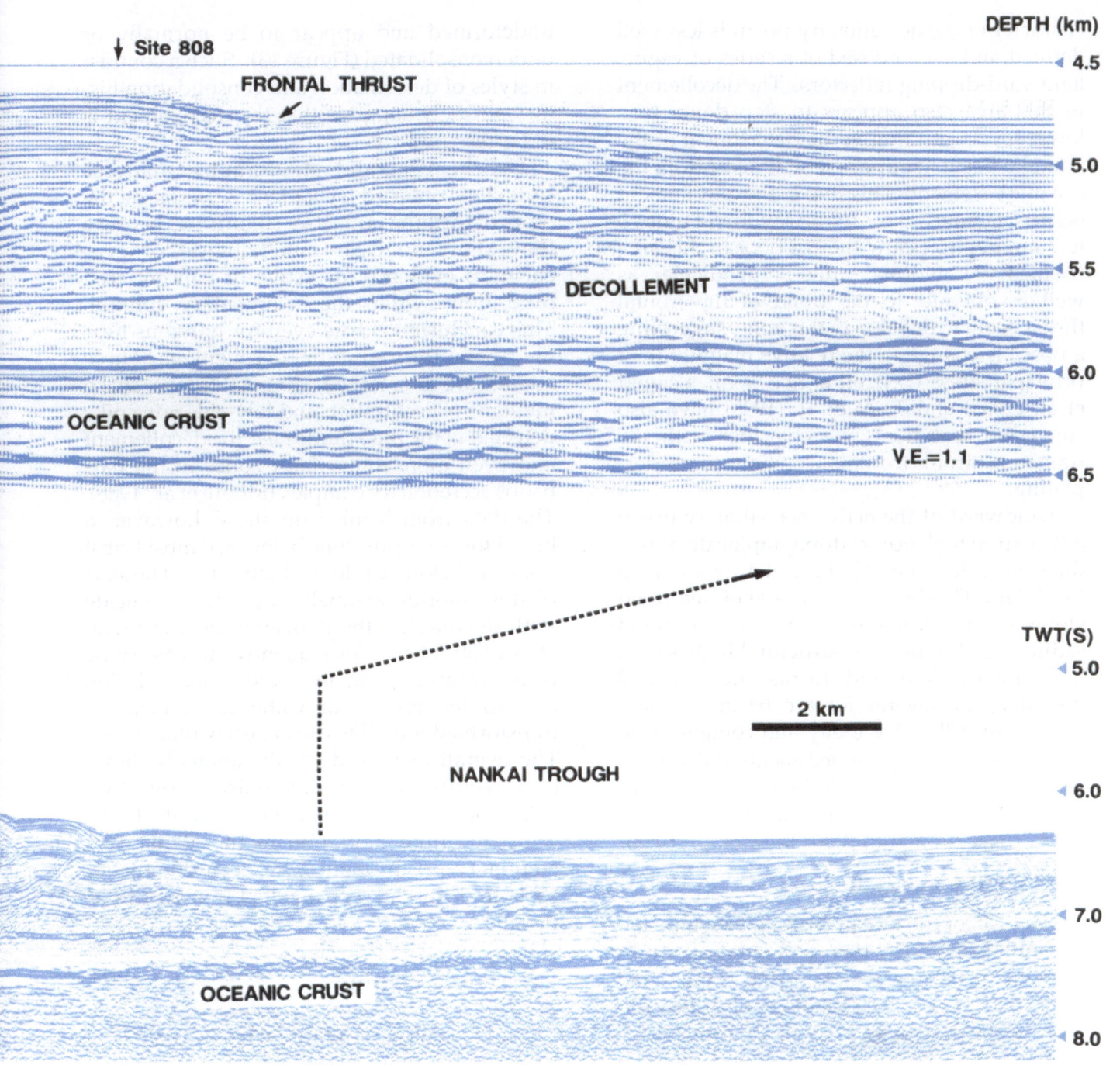

Figure 11   Multichannel seismic profile at the toe of the Nankai accretionary prism and its interpretation. The location of ODP site 808 (Leg 131) is also shown (Shipboard Scientific Party, ODP Leg 131, 1990).

and folding as deduced from the structural studies of the Shimanto Belt.

The landward dipping faults converge down-section to a detachment zone (or décollement) which, in some seismic lines, can be traced to about 30 km landward of the trench area (Moore et al., 1990). The décollement zone occurs at a well-defined horizon in seismic reflection profiles and can be placed within the upper section of the Shikoku Basin hemipelagic sediments. Core descriptions from Leg 131 of the Ocean Drilling Program revealed a 19-m-thick shear zone with abundant polished breccia or scaly clay (Taira et al., 1991).

Landward of the frontal thrust, the seismic

character of the accretionary prism is less well defined and is composed of a series of vague, landward-dipping reflectors. The décollement in this area also appears to step down to a lower level, resulting in the detachment of the Shikoku Basin hemipelagic sediments from the oceanic basement. This step in the décollement occurs at about 4.5 sec (two-way-travel time, 5 to 7 km depth) below the sea floor in this area (Ashi, 1991). Normal faults perpendicular, as well as oblique, to the trend of the frontal thrust begin to develop in this zone, suggesting a possible change in the relative magnitude of principal stress (Leggett et al., 1985). Leggett et al. (1985) and Taira et al. (1988) have also suggested that this zone coincides with the transition from frontal accretion to under-plating.

Landward of the active accretionary prism a structural high occurs (topographically this is the trench slope break) at a water depth of 0.5 to 1.5 km. This high is composed of deformed slope basin sediments and older accreted sediments. Locally, this structural high acts as a sediment dam and forms the seaward boundary of several forearc basins. These basins are still active today and contain up to 1 or 2 km of Neogene sediments (Okamura, 1990). The basins are underlain by older accretionary prism material, probably equivalent to the Tertiary Shimanto Belt (Taira et al., 1988).

9-2 ODP Leg 131 Results

Leg 131 of the Ocean Drilling Program successfully penetrated 1327 m of the stratigraphy present in the toe region of the Nankai accretionary prism. The lithology of drilled section is illustrated in a series of core photos (Photos 9-2-1 to 9-2-16). Sampled intervals included the frontal thrust, the décollement and part of the approximately 15 Ma oceanic crust that is being subducted beneath southwest Japan (Taira et al., 1991). One of the most exciting and spectacular results of the Nankai drilling project, however, was the discovery of a sharp contrast in the intensity of deformation above and below the décollement (Figures 12 and 13). Above the décollement the sediments are highly deformed and apparently overconsolidated. In contrast, below the décollement the sediments are virtually undeformed and appear to be normally or underconsolidated (Figure 13). Such a contrast in styles of deformation and consolidation history strongly suggests that the décollement is overpressured.

Geochemical and stratigraphic analyses completed onboard the *JOIDES Resolution* at the time of drilling are also significant because they show no clear mineralogical or geochemical evidence for active or concentrated fluid flow within the sedimentary column. This conclusion is true even for horizons near the frontal thrust and the décollement. These results are in marked contrast to the sharp methane concentration and low-chloride spike detected in the pore fluids near the décollement and even more minor fault zones in the Barbados accretionary complex (Moore et al., 1988). The data from Nankai do show, however, a broad low-chloride zone below 550 mbsf (mbsf = meters below sea floor) (Figure 13). The start of this chloride anomaly appears to coincide with the onset of the illitization of smectite at about 530 mbsf. Consequently, at least some of the chloride anomaly could reflect dilution due to the release of water as smectite is transformed into illite during early diagenesis. The overall magnitude of the anomaly, however, requires at least some pulse of low chloride fluid flow in the past (Taira et al., 1991). Thus, the fluid migration in the Nankai accretionary prism might have been dominated by both diffuse and channelled transient fluid flow.

9-3 Fluid Venting

Fluid venting and associated biological communities have been observed at several locations along the Nankai Trough and forearc through submersible dives, deep-sea photography and dredging. The best-studied areas are located in (1) the eastern Nankai Trough (and east of Site 808), where the KAIKO and KAIKO-Nankai expeditions revealed an extensive distribution of biological communities (Le Pichon et al., 1987; Kobayashi et al., 1989; Le Pichon et al., 1990) and (2) Sagami Trough (Hashimoto, 1989). The communities occur basically in two regions: in the frontal part of the accretionary prism and along the upper

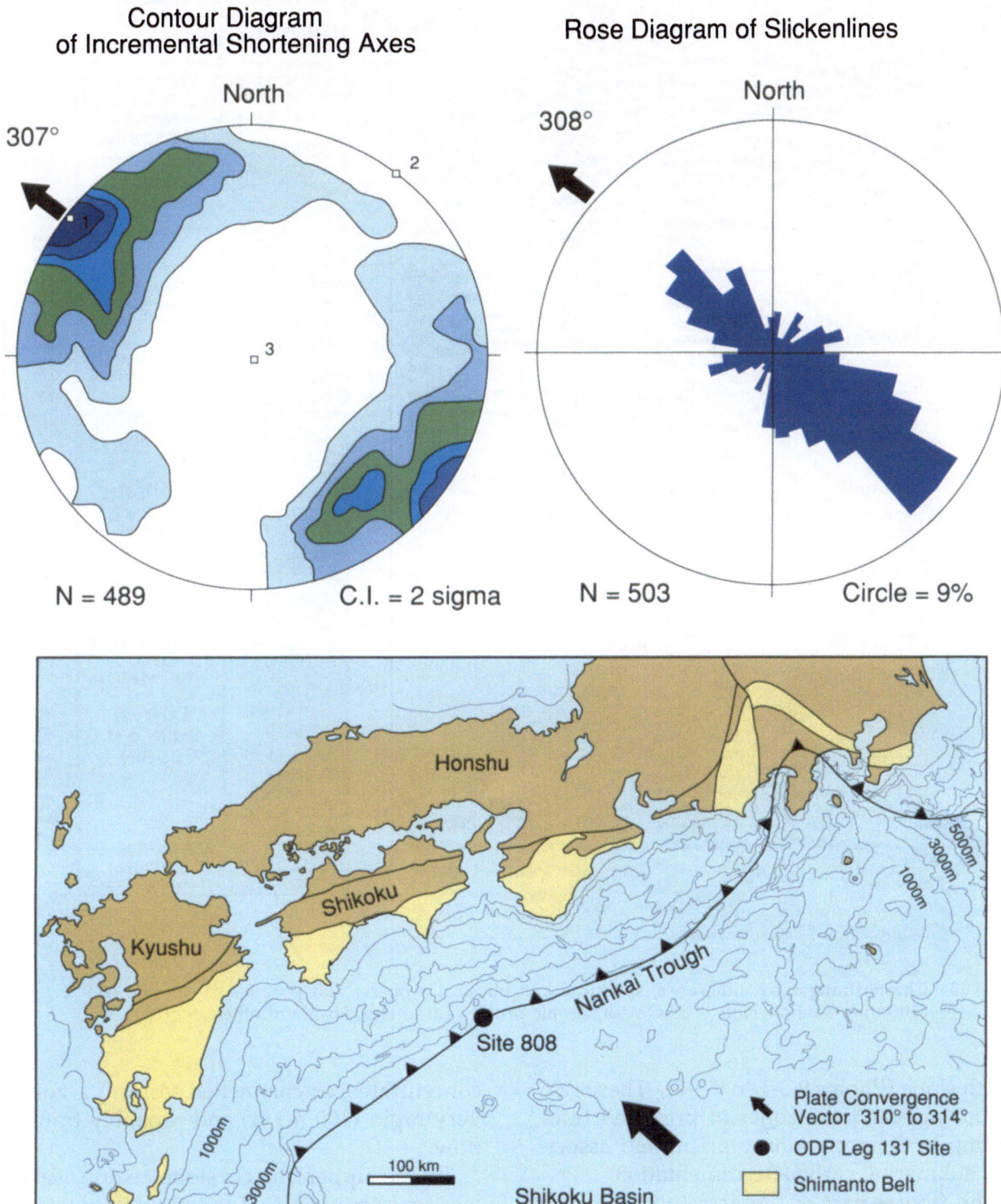

Figure 12   Diagrams showing the structural data and orientation of compressional direction obtained from ODP Leg 131 core analysis. The contour diagram shows the distribution of shortening axes deduced from faults and deformation bands (shear bands) plotted on the lower hemisphere of stereonet. The rose diagram to the left shows the distribution of slickenlines on fault surfaces. These two diagrams indicate 307-308 degree directions of shortening. The map shows the location of Site 808 and direction of plate convergence vector obtained from seismic mechanism analysis. Site 808 structural data provided for the first time a firm documentation of correlation of core-scale structural data to the regional scale plate convergence direction.

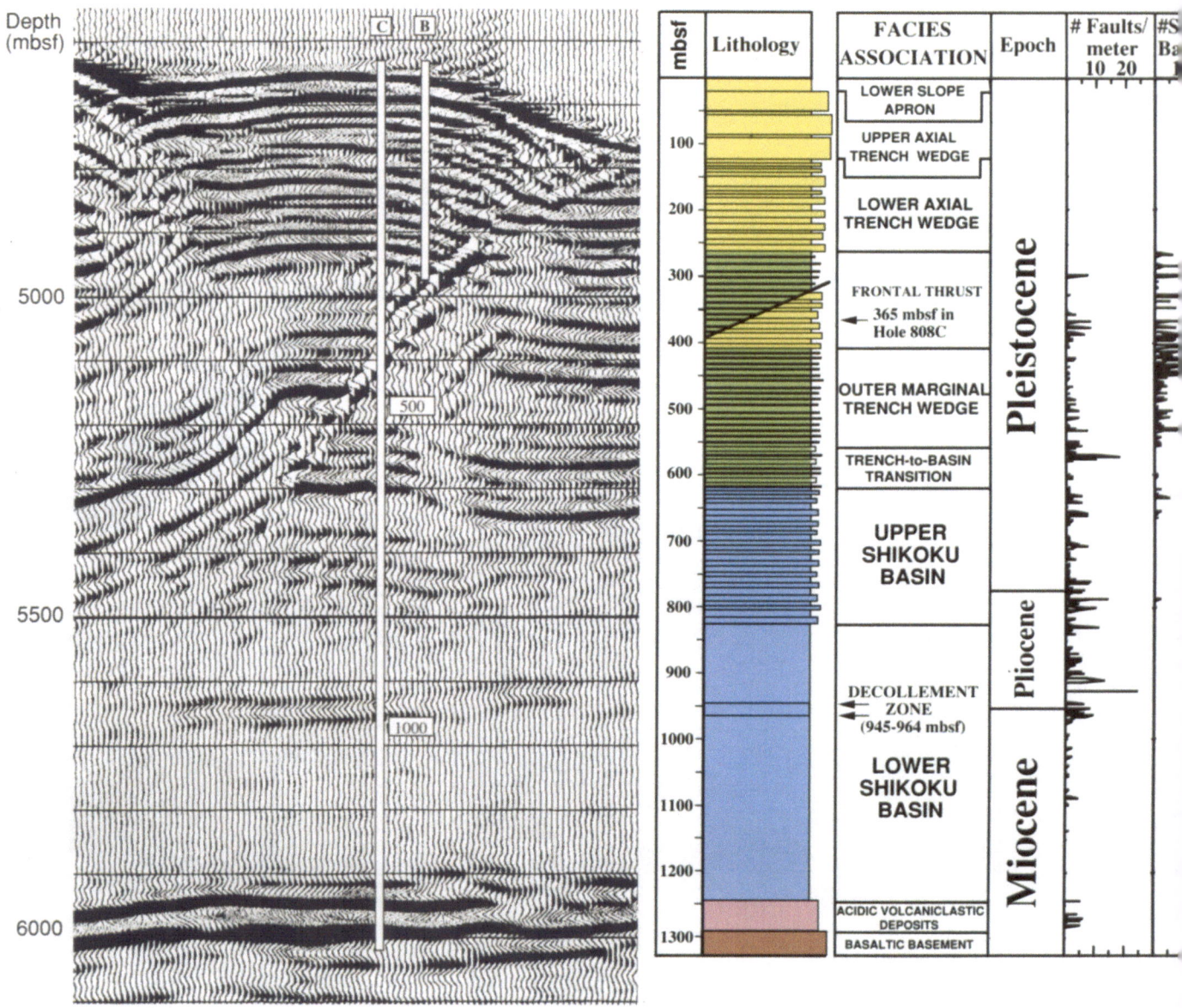

Figure 13   Chronostratigraphy, lithofacies, structural and physical property and pore water chloride concentration data of Hole 808B,C with seismic profile (Taira, Hill, Firth and others, 1991).

trench slope (Photos 9-3-1 to 9-3-7). These two regions appear to be zones of preferred fluid venting in the accretionary prism and associated submarine carbonate cementation.

The frontal part of the accretionary prism also seems to be a preferred zone of fluid expulsion in other submarine accretionary prisms (Moore et al., 1988, 1990). These sites have been interpreted to be loci of rapid tectonic dewatering and/or zones of fluid flow that have been channelled up-section from the décollement (Moore et al., 1990). Henry et al. (1990) have also suggested that fluids are

concentrated or channelled to form a zone of very rapid (100 m/yr) and probably transient flow.

In the upper trench slope region, several locations are identified as sites of fluid venting, including both the eastern (Kobayashi et al., 1989; Le Pichon et al., 1990) and western Nankai Trough regions (Okamura et al., 1986). The occurrence of biological communities in these regions seems to be associated with zones of active faulting and erosion, indicating that the accretionary prism is still actively deforming and dewatering in this region.

sity (%)

40 60

Cl (mM)

500 550

9-2-1: Slump deposits found in the uppermost part of Site 808. Note the recumbent nature of the fold within unlithified sediments. ODP Leg 131, Hole 808A, Core 9H, Section 3, 91-109 cm (73 mbsf, mbsf = meter below seafloor). Nankai Trough (16-Figure 4: location is the same for 9-2-1 to 9-2-15).

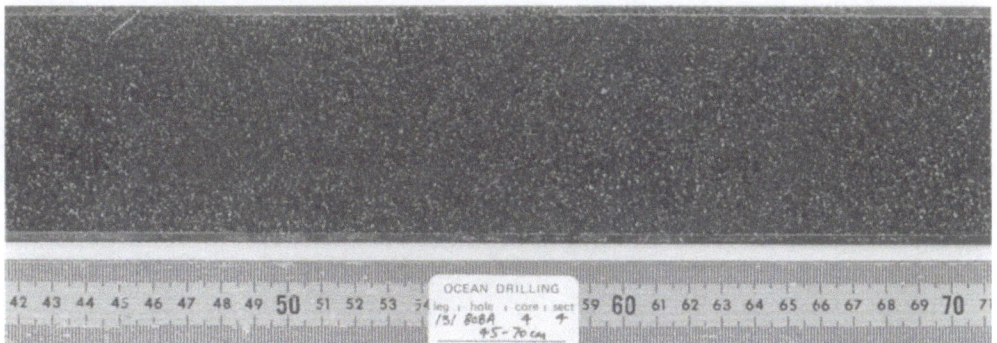

9-2-2: Medium to coarse-grained massive bedded sandy turbidite (Upper axial channel facies). ODP Leg 131, Hole 808A, Core 4H, Section 4, 40-72 cm (31 mbsf).

9-2-3: Fractured mudstone within the shear zone of the frontal thrust. ODP Leg 131, Hole 808C, Core 8, Section 1, 6-30 cm (376 mbsf).

9-2-4:   Shear band offset by low angle reverse faults in mudstone bed. ODP Leg 131, 808C, Core 15, Section 2, 15-24 cm (436 mbsf).

9-2-5:   Conjugate shear bands in mudstone bed at Hole 808C, Core 16, Section 2, 17-26 cm (445 mbsf).

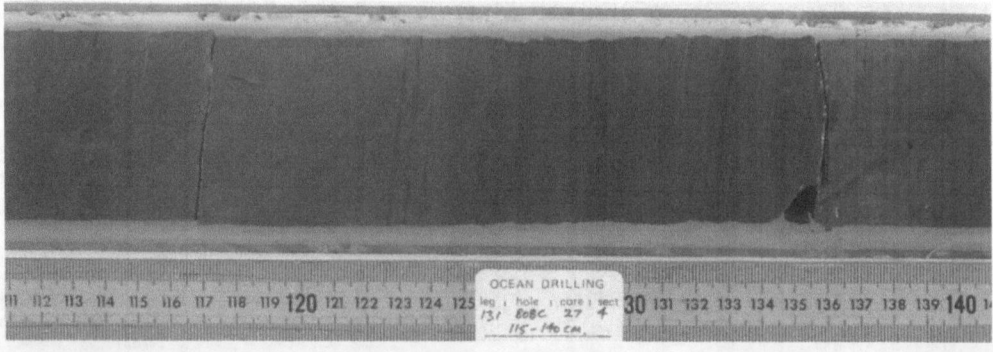

9-2-6:   Current ripples in fine-grained sandy turbidite. ODP Leg 131, Hole 808C, Core 24, Section 4, 111-141 cm (526 mbsf).

9-2-7:   A high-angle reverse faults. ODP Leg 131, Hole 808C, Core 25, Section 1, 65-81 cm (531 mbsf).

9-2-8:   Volcanic ashes intercalated with Upper Shikoku Basin facies. Note the burrowed upper part of each layer. ODP Leg 131, Hole 808C, Core 32, Section 6, 120-149 cm (605 mbsf).

9-2-9:   A sand dike found in mudstone of the Lower Shikoku Basin facies. ODP Leg 131, Hole 808C, Core 54, Section 4, 18-29 cm (814 mbsf).

9-2-10: Slicken lines on a fault surface in mudstone bed. ODP Leg 131, Hole 808C, Core 64, Section 4 (909 mbsf).

9-2-11: Fractured mudstone within the décollement zone. Note the polished surface of mudstone. ODP Leg 131, Hole 808C, Core 69, Section 4, 97-117 cm (959 mbsf).

9-2-12: Fractured mudstone recovered from the décollement zone. ODP Leg 131, Hole 808C (960 mbsf). Note the anastomosing fracture pattern.

9-2-13: *Zoophycus* bioturbated mudstone representing a typical lithology of Lower Shikoku Basin facies. ODP Leg 131, Hole 808C, Core 73, Section 1 (993 mbsf).

9-2-14: Greenish-colored altered acidic tuff beds overlying the basement. These thick ash beds seem to be related to forearc igneous activity associated with active spreading center subduction during the Middle Miocene. ODP Leg 131, Hole 808C, Core 101, Section 4 (1257 mbsf).

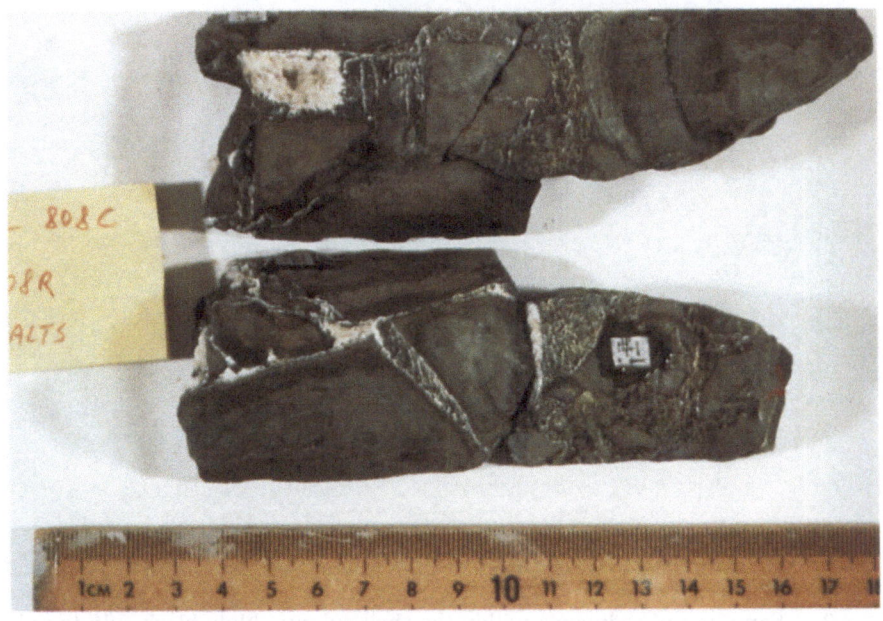

9-2-15: Sectioned core surfaces of pillow breccia and hyaloclastite composing the uppermost part of igneous basement of the Shikoku Basin. ODP Leg 131, Hole 808C, Core 108 (1318 mbsf).

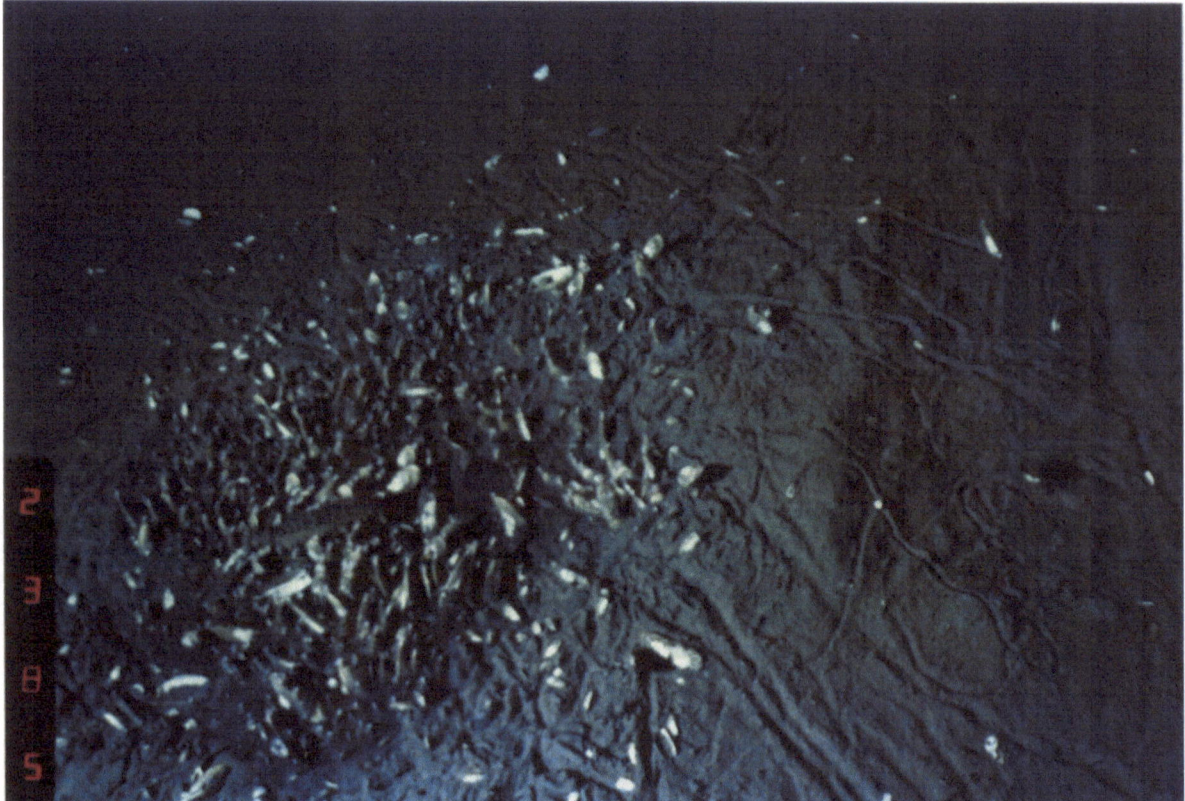

9-3-1:   Highly concentrated circular *Calyptogena* shell colony at the eastern Nankai Trough. Note the near-surface sulfide reduction zone indicated by black sediments within the colony. Water depth 3785 m. Photo taken by the submersible *Nautile*, KAIKO-Nankai Project. Eastern Nankai Trough (15-Figure 4).

9-3-2:   Sampling of sediments within the shell colony. Note black sulfide reduced sediments in box corer. Water depth 3830 m. Photo taken by the submersible *Nautile*, KAIKO Project. Eastern Nankai Trough (16-Figure 4).

9-3-3:   Close-up view of the shell communities at the Sagami Trough. Water depth 1985 m. Photo taken by the submersible *Shinkai 2000*. Courtesy by Japan Marine Science and Technology Center (JAMSTEC). Eastern Nankai Trough (15-Figure 4).

9-3-4: Dispersed occurrence of living shells. Water depth 3780 m. Photo taken by the submersible *Nautile*, KAIKO-Nankai Project. Eastern Nankai Trough (15-Figure 4).

9-3-5: Scattered shells and tube worms. Water depth 3836 m. Photo taken by the submersible *Nautile*, KAIKO Project. Eastern Nankai Trough (16-Figure 4).

9-3-6: Submarine carbonated cemented outcrops near active vents of upper slope region. Successive cementation, erosion and recementation were responsible for forming these complexes near vent fossil shell beds. Water depth 2183 m. Photo taken by the submersible *Nautile*, KAIKO-Nankai Project. Eastern Nankai Trough (15-Figure 4).

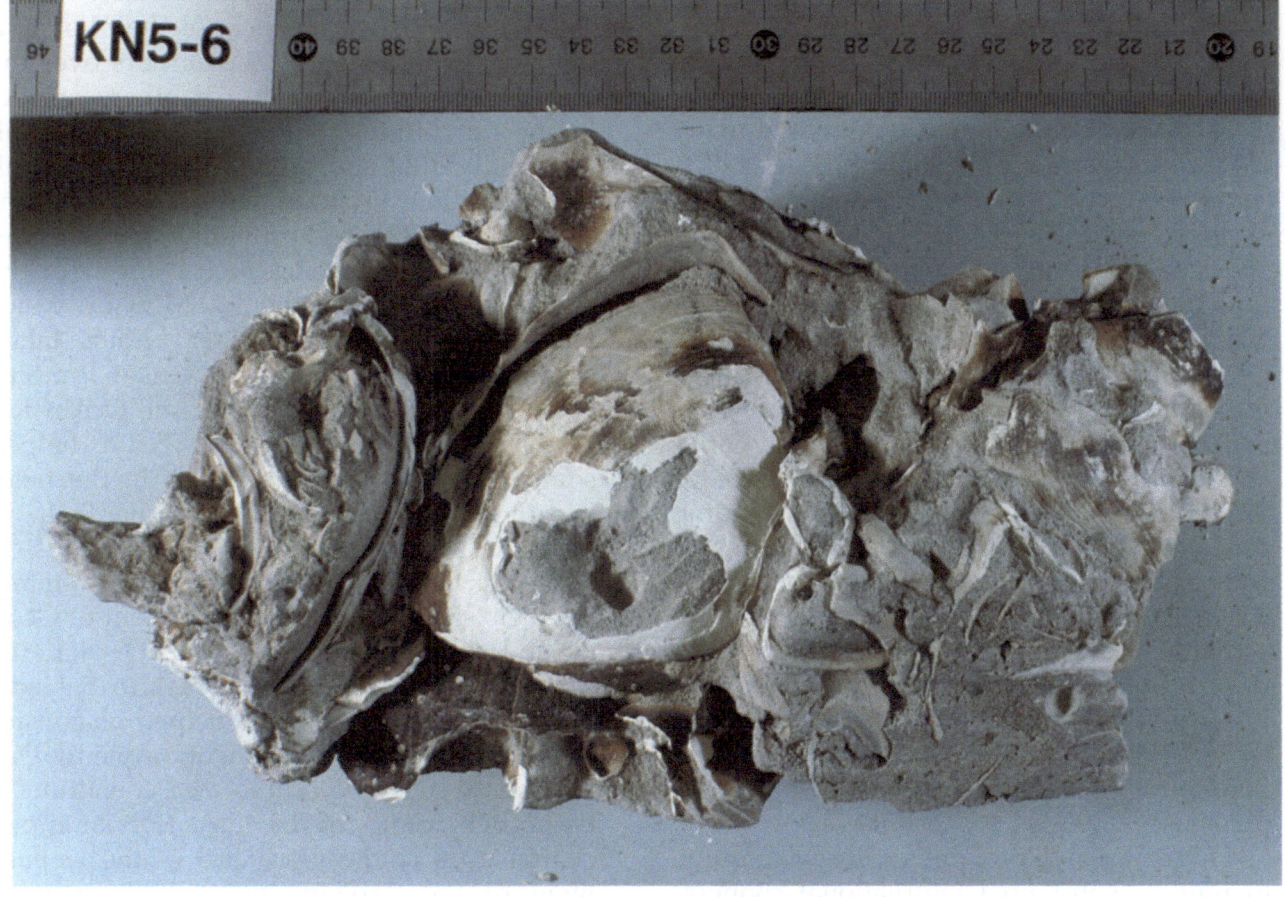

9-3-7: Carbonate cemented fossil shell beds sampled by submersible. Shells are only 15,000 yrs old. Water depth 1978 m. Sample taken by the submersible *Nautile*, KAIKO-Nankai Project. Eastern Nankai Trough (15-Figure 4).

10 CONCLUSIONS

Understanding the processes of accretion at convergent plate boundaries is a fundamental research objective in the earth sciences. This area of research contributes substantially to our understanding of the evolution of mountain belts and provides important insights into the origin of the continents and continental crust. The Japanese Islands, for example, have evolved primarily as deep-sea sediments and pieces of oceanic crust were accreted to the continental margin of Asia, causing the margin to grow over time. In fact, similar processes of accretion, deformation and metamorphism have been documented for the growth of most of Asia, including the Tibetan Plateau, as well as for the Cordillera of western North America. Recent work in the Early Archean rocks of western Australia also suggests that accretionary processes may have been occurring very early in the history of the earth (e.g., as early as 3.0 b.y. ago) (personal communication from S. Kiyokawa). Consequently, many of the processes we have highlighted in this photographic atlas are as fundamental to understanding the earth's evolution as they are to understanding the world we live in.

We have also attempted to highlight many of the more specific geologic processes that characterize convergent plate boundaries. For example, recent research programs have emphasized the importance of fluids and fluid flow in these tectonic environments. As mentioned in the last few chapters of this atlas, these studies have shown that the chemical, mechanical and thermal architectures of submarine accretionary prisms are strongly influenced by the presence of fluids. The source for these fluids, however, remains an outstanding problem. The fluids may originate from seawater that is incorporated with the sediment as it is accreted or from the dewatering of clays as they are underthrust and metamorphosed. Future research directions will therefore probably focus on documenting the source of these fluids as well as on the relation between fluids, fluid flow, deformation and metamorphism.

Future research programs will also continue to rely on substantial international cooperation, because convergent plate boundaries, by their very nature, cross political boundaries.

Moreover, the facilities that are needed to effectively study the processes occurring along these boundaries are often expensive, or one-of-a-kind. Cooperative international programs would therefore allow teams of scientists from different fields and different countries to achieve a common goal – to better understand one of the Earth's most dynamic environments – subduction zones.

Acknowledgments

We express our appreciation to Mr. W. Izumi, Ms. S. Schmidt, Ms. M. Shimizu and Ms. M. Komatsu of University of Tokyo Press and Mr. H. Adachi of Heihan Printing for their cooperation and encouragement in publishing this atlas. A Grant-in-aid for the Publication of Scientific Research Results from the Ministry of Education, Science and Culture (Monbusho) made this publication possible. T. Byrne also wishes to gratefully acknowledge the support of a Fellowship from the Japan Society for the Promotion of Science. Most of the photos in this atlas were taken by the authors over the past 15 years; the sources of others are acknowledged in the photo captions. We thank each individual and organization. We thank Ocean Drilling Program and shipboard scientists of Leg 131 for providing us the core photos. Also Ocean Research Institute of the University of Tokyo and IFREMER of France provided us the under-water photos taken during KAIKO and KAIKO-Nankai Projects. S. Morita and S. Kiyokawa of the Ocean Research Institute, University of Tokyo helped some of the photographic work in the field. We thank Ms. T. Kanehara for her help in computer artwork.

In the field, we were supported by many people. The staffs of the Department of Geology of Kochi University, especially Drs. J. Katto, M. Okamura, K. Kodama, M. Tashiro, S. Hada, T. Suzuki and H. Ishizuka, helped us considerably as co-workers. The work done by Drs. J. Hibbard and L. DiTullio also constitutes a basic background for the atlas. Special appreciation goes to the graduates of the Department of the Geology of Kochi University, without whose support it would have been impossible to carry out the investigation in the

Shimanto Belt. A number of residents of the city of Muroto have helped us considerably logistically. The assistance of Ikuko Taira during the field work was much appreciated.

Finally, we thank Mother Nature for creat-ing the fascinating geology of the beautiful coastal area of the island of Shikoku, and hope that these exposures will be preserved for years to come.

REFERENCES

Agar, S.M., Cliff, R.A., Duddy, I.R., and Rex, D.C., 1989. Accretion and uplift in the Shimanto Belt, SW Japan. Journal of Geological Society, London, v.146, pp.893-896.

Angelier, J., and Huchon, P., 1987. Tectonic record of convergence changes in a collision area: The Boso and Miura peninsulas, Central Japan. Earth and Planetary Science Letter, v.81, pp.397-408.

Ashi, J., 1991. Structure and hydrogeology of the Nankai accretionary prism. D.Sc. thesis, Department of Geology, University of Tokyo.

Byrne, T., and Fisher, D., 1990. Evidence for a weak and overpressured decollement beneath sediment-dominated accretionary prisms. Journal of Geophysical Research, v.95, pp.9081-9097.

Chamot-Rooke, N., Renard, V., and Le Pichon, X., 1987. Magnetic anomalies in the Shikoku Basin: A new interpretation. Earth and Planetary Science Letter, v.83, pp.214-228.

DiTullio, L.D., and Byrne, T., 1991. Deformation in the shallow levels of an accretionary prism: The Eocene Shimanto Belt of southwest Japan. Geological Society of America Bulletin, v.102, pp.1420-1438.

DiTullio, L.D., Laughland, M., and Byrne, T., in press. Thermal history and constraints on deformation from illite crystallinity and vitrinite reflectance in the shallow levels of an accretionary prism: Eocene Shimanto Belt, southwest Japan. Geological Society of America, Special Paper.

Eto, T., Oda, M., Hasegawa, S., Honda, N., and Funayama, M., 1987. Geologic age and paleoenvironment based upon microfossils of the Cenozoic sequence in the middle and northern parts of the Miura Peninsula. Science Report of Yokohama National University, Series 2, v.34, pp.41-57 (in Japanese with English abstract).

Fisher, D., and Byrne, T., 1987. Structural evolution of underthrusted sediments: Evidence from Kodiak Island, Alaska. Tectonics, v.6, pp.775-793.

Fryer, P., Pearce, J.A., Stokking, L.B., and others, 1990. Proceedings of the Ocean Drilling Program, Initial Reports., v.125. College Station, TX (Ocean Drilling Program), 1092 p.

Hasebe, N., Tagami, T., and Nishimura, S., in press. Evolution of the Shimanto accretionary complex: A fission-track thermochronologic study. Geological Society of America, Special Paper.

Hashimoto, J., Ohta, S., Tanaka, T., Hotta, H., Matsuzawa, S., and Sakai, H. 1989. Deep-sea communities dominated by the giant clam, Calyptogena soyoae, along the slope foot of Hatsushima Island, Sagami Bay, central Japan. Paleogeography Paleoclimatology Paleoecology, v.71, pp.179-192.

Henry, P., Le Pichon, X., Lallemant, S., Foucher, J., Westbrook, G., and Horbart, M., 1990. Mud volcano field seaward of the Barbados accretionary complex: A deep-towed side-scan sonar survey. Journal of Geophysical Research, v.95, pp.8917-8929.

Hibbard, J.P., and Karig, D.E., 1987. Sheath-like folds and progressive fold deformation in Tertiary sedimentary rocks of the Shimanto accretionary complex, Japan. Journal of Structural Geology, v.9, pp.845-857.

Hibbard, J.P., and Karig, D.E., 1990. Structural and magmatic responses to spreading ridge subduction: An example from southwest Japan. Tectonics, v.9, pp.207-230.

Ishikawa, T., 1982. Radiolarians from the southern Shimanto Belt (Tertiary) in Kochi Prefecture, Japan. News of Osaka Micropaleontologies, Special Volume, v.5, pp.399-407.

Isozaki, Y., Maruyama, S., and Furuoka, F., 1990. Accreted oceanic materials in Japan. Tectonophysics, v.181, pp.179-205.

Isozaki, Y., and Itaya, T., 1991. Pre-Jurassic klippe in northern Chichibu Belt in west-central Shikoku, Southwest Japan: Kurosegawa Terrane as a tectonic outlier of the pre-Jurassic rocks of the Inner Zone. Journal of Geological Society of Japan, v.97, pp.431-450.

Kagami, H., Karig, D.E., Coulbourn, W., and others, 1986. Initial Reports of the Deep Sea Drilling Project. Washington D.C., U.S. Government Printing Office, v.87, 985 p.

Kakimi, T., Hirayama, J., and Kageyama, K., 1966. Tectonic stress-fields deduced from the minor fault systems in the northern part of the Miura Peninsula. Journal of Geological Society of Japan, v.72, pp.469-489.

Kano, K., Nakaji, M., and Takeuchi, S., 1991. Asymmetrical melange fabrics as possible indicators of convergent direction of plates: A case study from the Shimanto Belt of the Akaishi Mountains, central Japan. Tectonophysics, v.185, pp.375-378.

Karig, D.E., and Angevine, C.L., 1986. Geologic constraints on subduction rates in the Nankai Trough, in Kagami, H., Karig, D.E., and others, eds., Initial Reports of the Deep Sea Drilling Project. Washington D.C., U.S. Government Printing Office, v. 87, pp.789-796.

Karig, D.E., and Lundberg, N., 1990. Deformation bands from the toe of the Nankai accretionary prism. Journal of Geophysical Research, v.95, pp.9099-9109.

Katto, J., 1969. Geology of Kochi Prefecture. Kochi Municipal Library Press, Kochi City, 316 p.

Katto, J., and Taira, A., 1979. The Misaki Group (Miocene), southwestern Shikoku. Research Reports of Kochi University, v.15, pp.57-63 (in Japanese with English abstract).

Kimura, G., Koga, K., and Fujioka, K., 1989. Deformed soft sediments at the junction between the Mariana and Yap Trenches. Journal of Structural Geology, v.11, pp.463-472.

Kimura, G., and Mukai, A., 1991.Underplated units in an accretionary complex: Melange of the Shimanto Belt of Eastern Shikoku, Southwest Japan. Tectonics, v.10, pp.31-50.

Knipe, R.J., 1986. Microstructural evolution of vein arrays preserved in Deep Sea Drilling Project cores from the Japan Trench, Leg 57, in Moore J.C., ed., Structural Fabrics in Deep Sea Drilling Cores from Forearcs. Geological Society of America Memoir 166, pp.75-87.

Kobayashi, K., Le Pichon, X., and others, 1989. Quantitative evaluation of large-scale fluid venting on the frontal portion of the eastern Nankai accretionary prism using the 6000 m Nautile submersible. EOS, v.70, p.1381.

Kobayashi, K., and Nakada, M., 1978. Magnetic anomalies and tectonic evolution of the Shikoku interarc basin. Journal of Physics of the Earth, v.26, pp.391-402.

Kodama, K., Fukui, H., and Muro-oka, K., 1988. Kite aerial photography and its application to geology. Journal of Geological Society of Japan, v.5, pp.381-385.

Komatsu, M., Shiraishi, R., Matsuo, M., and Tanaka, R., 1991. High T/P metamorphism in the lower part of the Shimanto Belt during middle Miocene time [abs.]: Geological Society of Japan, Annual Meeting, 98th, Matsuyama, 1991. pp.34-35 (in Japanese).

Kumon, F., Suzuki, H., Nakazawa, K., Tokuoka, T., Harata, T., Kimura, K., Nakaya, S., Ishigami, T., and Nakamura, K., 1988. Shimanto Belt in the Kii Peninsula, Southwest Japan. Marine Geology, v.12, pp.71-96.

Laughland, M.M., and Underwood, M.B., in press. Vitrinite reflectance and estimates of paleotemperature within the upper Shimanto Belt, Muroto Peninsula, Shikoku, Japan. Geol. Soc. America Special Paper.

Leggett, J. K., Aoki, Y., and Toba, T., 1985. Transition from frontal accretion to underplating in a part of the Nankai Trough accretionary complex off Shikoku (SW Japan) and extensional features on the lower trench slope. Marine Petroleum Geology, v.2, pp.131-141.

Leggett, J.K., Lundberg, N., Bray, C.J., Cadet, J.P., Karig, D.E., Knipe, R.J., and von Huene, R., 1987. Extensional tectonics in the Honshu forearc, Japan: Integrated results of DSDP Legs 57, 87 and reprocessed multichannel seismic refraction profiles. Continental Tectonics: Geological Society of America, Special Publication, v.28, pp.593-609.

Le Pichon, X., Iiyama, T., Chamley, H., Charvet, J., Faure, M., Fujimoto, H., Furuta, T., Ida, Y., Kagami, H., Lallemant, S., Leggett, J., Murata, A., Okada, H., Rangin, C., Renard, V., Taira, A. and Tokuyama, H., 1987. Nankai Trough and the fossil Shikoku Ridge, results of Box 6 Kaiko survey. Earth and Planetary Science Letter, v.83,

pp.186-198.

Le Pichon, X., Foucher, J., Boulegue, J., Henry, P., Lallemant, S., Benedetti, M., Avedik, F., and Mariptt, A., 1990. Mud volcano field seaward of the Barbados accretionary complex: A submersible survey. Journal of Geophysical Research, v.95, pp.8931-8944.

Lundberg, N., and Moore, J.C., 1986. Macroscopic structural features in Deep Sea Drilling Project cores from forearc regions, in Moore, J.C., eds., Structural Fabrics in Deep Sea Drilling Project Cores from Forearcs. Geological Society of America Memoir 166, pp.13-44.

Mackenzie, J.S., 1991. Accretionary related deformation in the Shimanto Belt of eastern Kyushu, S.W. Japan. Ph.D. thesis, University of London.

Matsumoto, E., and Hirata, M., 1972. *Akebiconcha uchimuraensis* (Kuroda) from the Oligocene formations of the Shimanto terrain. Bulletin of National Science Museum, Tokyo, v.15, pp.753-762.

Miyake, Y., 1988. Petrology of the Shionomisaki igneous complex, southwest Japan and its implication for the ophiolite generation. Modern Geology, v.12, pp.283-302.

Moore, J.C., and Scientific Party of ODP Leg 110, 1988. Tectonics and hydrogeology of the northern Barbados: Results from Ocean Drilling Program leg 110. Geological Society of America Bulletin, v.100, pp.1578-1593.

Moore, G.F., Shipley, T.H., Stoffa, P.L., Karig, D.E., Taira, A., Kuramoto, S., Tokuyama, H., and Suyehiro, K., 1990. Structure of the Nankai Trough accretionary zone from multichannel seismic reflection data. Journal of Geophysical Research, v.95, pp.8753-8765.

Niitsuma, N., Matsushima, Y., and Hirata, D., 1989. Abyssal molluscan colony of Calyptogena in the Pliocene strata of the Miura Peninsula, central Japan. Paleogeography, Paleoclimatology and Paleoecology, v.71, pp.193-203.

Ogawa, Y., 1980. Beard-like veinlet structure as fracture cleavage in the Neogene siltstone in the Miura and Boso Peninsulas, central Japan. Science Reports Department Geology Kyushu University, v.13, pp.321-327 (in Japanese with English abstract).

Ogawa, Y., and Taniguchi, H., 1989. Origin and emplacement of basaltic rocks in the accretionary complexes in SW Japan. Offioliti, v.14, pp.177-193.

Okamura, M., and Uto, H., 1982. Notes on stratigraphic distributions of radiolarians from the Lower Cretaceous sequence of chert in the Yokonami Melange of Shimanto Belt, Kochi Prefecture, Shikoku. Res. Repts. of Kochi Univ., v.31, pp.87-94.

Okamura, M., and Taira, A., 1984. Geology and paleontology of the southern Shimanto Belt (Tertiary) in the Muroto Peninsula, in Saito, T., and Okada, H., eds., Biostratigraphy and international correlation of the Paleogene system in Japan. Yamagata University, pp.81-83.

Okamura, Y., Tanaka, T., and Nakamura, K., 1986. Diving survey of the knolls on the trench slope break off Kochi, Southwest Japan. Technical Reports of Japan Marine Science and Technology Center (JAMSTEC), Special Issue on the 2nd Symposium on Deep Sea Research using the Submersible "SHINKAI 2000" System, pp.173-189 (in Japanese with English abstract).

Okamura, Y., 1990. Geologic structure of the upper continental slope off Shikoku and Quaternary tectonic movement of the outer zone of southwest Japan. Journal of Geological Society of Japan, v.96, pp.223-237 (in Japanese with English abstract).

Pickering, K.T., Agar, S.M., and Prior, D.J., 1990. Vein structure and the role of pore fluids in early wet-sediment deformation, Late Miocene volcaniclastics, Miura Group, SE Japan, in Knipe, R.J., and Rutter, E.H., eds., Deformation Mechanics, Rheology and Tectonics. Geological Society of London Special Publication, v.54, pp.417-430.

Saito, T., 1980. An early Miocene (Aquitanian) planktonic foraminiferal fauna from the Tsuro Formation, the youngest part of the Shimanto Supergroup, Shikoku, Japan. in Taira, A., and Tashiro, M., eds., Geology and Paleontology of the Shimanto Belt: Selected. Papers. in Honor. of Prof. Jiro Katto. Rinyakosaikai Press, Kochi, pp.227-234.

Sakai, H., 1987. Storm barnacle beds and their deformation in the Murotomisaki olistostrome and melange complex, Shikoku. Journal of Geological Society of Japan, v.93, pp.617-620.

Seno, T., 1977. The instantaneous rotation vector of the Philippine Sea plate related to the Eurasian plate. Tectonophysics, v.102, pp.209-226.

Shih, T.C., 1980. Magnetic lineations in the Shikoku Basin, in Klein, G. deV., Kobayashi, K., and others, eds., Initial Reports of the Deep Sea Drilling Project. Washington D.C., U.S. Government Printing Office, v.58, pp.783-788.

Soh, W., Pickering, K,T., Taira, A., and Tokuyama, H., 1991. Basin evolution in the arc-arc Izu Collision Zone, Mio-Pliocene Miura Group, central Japan. Journal of Geological Society, London, v.148, pp.317-330.

Sugiyama, Y., 1989. Bend of the zonal structure of island arcs and oblique subduction as the cause of the bending. Bulletin of Geological Survey of Japan, v.40, pp.533-541 (in Japanese with English abstract).

Taira, A., Okamura, M., Katto, J., Tashiro, M., Saito, Y., Kodama, K., Hashimoto, M., Chiba, T., and Aoki, T., 1980. Lithofacies and geologic age relationship within melange zones of Northern Shimanto Belt (Cretaceous), Kochi Prefecture, Japan, in Taira, A., and Tashiro, M., eds., Geology and Paleontology of the Shimanto Belt: Selected Papers in Honor of Prof. Jiro Katto. Rinyakosaikai Press, Kochi, pp.179-214 (in Japanese with English abstract).

Taira, A., 1982. Paleotectonic setting of the Nagase and Kajisato Formations (Upper Cretaceous), Shikoku, in Matsumoto, T., and Tashiro, M., eds., Multidisciplinary Research in the Upper Cretaceous of the Monobe Área, Shikoku. Paleontological Society of Japan Special Paper, v.25, pp.15-25.

Taira, A., 1985. Sedimentary evolution of Shikoku subduction zone: The Shimanto Belt and Nankai Trough, in Nasu, N., and others, eds., Formation of Active Ocean Margins. Terra Scientific Publishing Company, Tokyo, pp.835-851.

Taira, A., and Niitsuma, N., 1986. Turbidite sedimentation in the Nankai Trough as interpreted from magnetic fabric, grain size, and detrital modal analyses, in Kagami, H., Karig, D.E., Coulbourn, W.C., and others, eds., Initial Reports of the Deep Sea Drilling Project. Washington D.C., U.S. Government Printing Office, v.87, pp.611-632.

Taira, A., Katto, J., Tashiro, M., Okamura, M., and Kodama, K., 1988. The Shimanto Belt in Shikoku, Japan: Evolution of Cretaceous to Miocene accretionary prism. Modern Geology, v.12, pp.5-46.

Taira, A., Tokuyama, H., and Soh, W., 1989. Accretion tectonics and evolution of Japan, in Ben-Avraham, Z., eds., The Evolution of the Pacific Ocean Margins. Oxford (Oxford University Press), pp.100-123.

Taira, A., Hill, I., Firth, J., and others, 1991. Proceedings of the Ocean Drilling Program, Initial Reports., v.131. College Station, TX (Ocean Drilling Program), 306 p.

Taylor, B., Fujioka, K., and others, 1990. Proceedings of the Ocean Drilling Program, Initial Reports., v.126. College Station, TX (Ocean Drilling Program), 1002 p.

Teraoka, Y., 1979. Provenance of the Shimanto geosynclinal sediments inferred from sandstone compositions. Journal of Geological Society of Japan, v.85, pp.753-769.

Tokunaga, T., in press. Two stage foldings developed in the Paleogene Shimanto Supergroup, southwestern Shikoku, Japan. Journal of Geological Society of Japan.

Underwood, M.B., Laughland, M.M., and Kang, S.M., in press. A comparison between organic and inorganic indicators of diagenesis and metamorphism, Upper Shimanto Group, Muroto Peninsula, Shikoku, Japan. Geol. Soc. America Special Paper.